Biology For Beginners

PROF. ELIZABETH WILFRED

Understanding The Basics Of Life

BIOLOGY FOR BEGINNERS

Copyright © Elizabeth Wilfred 2024

Copyright ©Elizabeth Wilfred 2024. All rights reserved.

No part of this book may be reproduced, distributed, or transmitted in any form or by any means, including photocopying, recording, or other electronic or mechanical methods, without the prior written permission of the author, except in the case of brief quotations embodied in critical reviews and certain other noncommercial uses permitted by copyright law.

FORWARD

Hey there!

Welcome to "Introduction To Biology," your ticket to exploring the wild world of life around us. I'm your guide through this jungle of knowledge. In these pages, we're diving deep into biology, tailor-made for high schoolers especially. From tiny cells to the giant ecosystems, we're unraveling the mysteries of life.

Biology isn't just facts—it's the key to unlocking the secrets of life itself. Together, we'll journey through evolution, marvel at the crazy variety of creatures, and ponder how everything's connected. With easy explanations, cool pictures, and real-life examples, I've whipped up this textbook to blow your mind. Whether you're into DNA drama, fascinated by funky animals, or curious about ecosystems, "Introduction To Biology" has got your back.

As you flip through these pages, don't be shy to ask questions and challenge ideas. Biology's always changing, so let's keep up! My wish is for this book to not only fill your brain with smarts but also light a fire for exploring nature's awesomeness. Let's go on this adventure together!

Happy exploring!
Amanda Smith

ACKNOWLEDGEMENT

I extend my heartfelt gratitude to everyone who contributed to the creation of "Introduction to Biology." This book has been a labor of love, and I am deeply thankful for the support and assistance I received along the way.

First and foremost, I want to thank my family for their unwavering encouragement and understanding during the writing process. Their belief in me kept me motivated even during the most challenging moments. I am immensely grateful to the educators and experts who generously shared their knowledge and insights, enriching the content of this textbook. Your expertise has helped shape the clarity and accuracy of the material presented.

I extend my appreciation to the editorial and production teams who worked tirelessly to bring this book to life. Your dedication to excellence ensured that every word, illustration, and layout element was polished to perfection. To my colleagues and peers who provided valuable feedback and encouragement throughout this journey, thank you for your support and camaraderie. Your insights and perspectives have been invaluable in refining the content and enhancing its relevance for students. I also want to acknowledge the students who will embark on this educational journey with "Introduction to Biology." Your curiosity and enthusiasm for learning inspire me, and it is my sincere hope that this textbook will serve as a valuable resource in your academic endeavors.

Last but not least, I want to express my gratitude to the readers of this book. Your interest in the field of biology fuels my passion for sharing knowledge, and I am honored to be a part of your learning experience.

Thank you all for being a part of this incredible journey.

TABLE OF CONTENT

FOREWORD 3
ACKNOWLEDGEMENT 4
CHAPTER ONE: INTRODUCTION TO BIOLOGY 1
   What Is Biology? 1
   Branches Of Biology 2
   Scientific Method in Biology 6
   Tools And Techniques In Biology 9
   Careers In Biology 12
CHAPTER TWO: THE STUDY OF CELLS 17
   Cell Theory 17
   Historical Development Of Cell Theory 18
   Cell Structure And Function 19
   Cell Membrane and Transport 33
   Cell Division: Mitosis and Meiosis 38
CHAPTER THREE: GENETICS AND HEREDITY **50**
   Mendelian Inheritance 53
   Genetic Variation 59
   DNA and Genes 61
   Protein Synthesis: Transcription And Translation 66
   Genetic Engineering And Biotechnology 68
CHAPTER FOUR: EVOLUTION AND NATURALSELECTION 75
   The Theory Of Evolution 71
   Evidence For Evolution 72

Mechanisms Of Evolution 74

Natural Selection 76

Speciation And Adaptive Radiation 78

Human Evolution 80

CHAPTER FIVE: BIODIVERSITY AND CLASSIFICATION 83

Introduction To Biodiversity 83

Classification and Taxonomy 86

Kingdoms Of Life 88

Plant Kingdom 126

Fungi, Protists, And Bacteria 141

Kingdom Protista 144

CHAPTER SIX: HUMAN ANATOMY AND PHYSIOLOGY 150

Introduction To The Human Body 150

Skeletal System 153

Muscular System 166

Nervous System 170

Circulatory System 174

Respiratory System 182

Digestive System 187

Excretory System 190

Reproductive System 192

Endocrine System 197

CHAPTER SEVEN: ECOLOGY AND ECOSYSTEMS 202

Introduction To Ecology And Ecosystem 202

Levels Of Organization In Ecology 204
Ecosystems And Biomes 205
Energy Flow In Ecosystems 225
Nutrient Cycling 229
Populations And Communities 244
Conservation Biology 245

CHAPTER EIGHT: VARIATION 248

Introduction To Variation 248
Types Of Variation 249
Sources Of Variation 251
Factors Influencing Phenotypic Variation 252

COMMON TERMS AND DEFINITIONS IN BIOLOGY 257

INTRODUCTION TO BIOLOGY

CHAPTER 1

■ What Is Biology?

Biology is the scientific study of life and living organisms. It is a branch of science that explores the structure, function, growth, evolution, and distribution of organisms on Earth. By examining the diverse forms of life, from the tiniest microorganisms to complex multicellular organisms, biology seeks to uncover the fundamental principles that govern the living world. At its core, biology seeks to answer questions about how life arose, how it functions, and how it interacts with the environment. It encompasses a wide range of disciplines, including genetics, physiology, ecology, microbiology, botany, zoology, and many more. These sub-disciplines allow for a deeper exploration of specific aspects of life, enabling scientists to specialize in their areas of interest while contributing to the broader understanding of biology as a whole.

The scope of biology extends beyond the study of individual organisms. It also encompasses the interactions between organisms and their environment, the processes of evolution and adaptation, and the intricate networks that connect different species in ecosystems. By studying the intricate web of life, biologists gain insights into the intricate balance that sustains life on our planet. Biology is a dynamic field that continually evolves as new discoveries are made and technologies advance. Through scientific inquiry and research, biologists uncover the mechanisms that underlie life's processes, from the molecular level to the ecosystem level. This knowledge forms the foundation for advancements in medicine, agriculture, conservation, and biotechnology.

One of the fundamental principles in biology is the concept of evolution. Evolutionary theory explains how species change over time, giving rise to the incredible diversity of life we observe today. Through the processes of natural selection, genetic variation, and speciation, organisms adapt to their environments, ensuring their survival and shaping the course of biological history. The study of biology is not only intellectually stimulating but also has practical applications that impact our daily lives. Biologists contribute to advancements in medical research, helping us understand and combat diseases. They study ecosystems to develop sustainable practices for conserving biodiversity and preserving natural resources. Biotechnology, a rapidly growing field within biology, utilizes living organisms and their components to develop innovative products and technologies that benefit society.

As we delve deeper into the study of biology, we begin to appreciate the interconnectedness of life and the remarkable complexity that exists within even the simplest organisms. From the intricate workings of a cell to the intricacies of ecosystems, biology allows us to unlock the mysteries of life and provides us with a deeper understanding of our place in the natural world.

■ Branches Of Biology

Biology, the scientific study of life and living organisms, encompasses numerous branches that focus on specific aspects of the subject. These branches collectively contribute to our understanding of the complexity and diversity of life on Earth. Here are some major branches of Biology:

Introduction To Biology

1. **Zoology:** Zoology is the branch of biology that deals with the study of animals. It involves the examination of animal behavior, anatomy, physiology, evolution, classification, and ecology. Zoologists study a wide range of organisms, from microscopic invertebrates to large vertebrates, and contribute to our understanding of animal life in its various forms.
2. **Botany:** Botany, also known as plant biology, is the branch of biology that focuses on the study of plants. Botanists explore plant structure, growth, reproduction, metabolism, and evolution. They investigate diverse plant groups, from algae and fungi to flowering plants, and their interactions with the environment.
3. **Microbiology:** Microbiology is the study of microorganisms, including bacteria, viruses, fungi, and protists. Microbiologists investigate the characteristics, behavior, genetics, and ecological roles of these microscopic organisms. They play a crucial role in understanding the impact of microorganisms on human health, disease, food production, and the environment.
4. **Genetics:** Genetics is the branch of biology concerned with the study of genes, heredity, and genetic variation. Geneticists explore the principles of inheritance, the structure and function of genes, and the mechanisms that regulate gene expression. This field has witnessed significant advancements in recent years with the advent of molecular genetics and genomics.
5. **Ecology:** Ecology is the study of the interactions between organisms and their environment. Ecologists investigate the relationships between living organisms, including their distribution, abundance, behavior, and adaptations. They explore ecological processes at various scales, from individual organisms to entire ecosystems, and contribute to our

understanding of biodiversity conservation and sustainable practices.
6. **Evolutionary Biology:** Evolutionary biology focuses on the study of the processes that drive the diversity and change of life forms over time. Evolutionary biologists examine the mechanisms of evolution, such as natural selection, genetic drift, and speciation. They explore the patterns and history of life on Earth, reconstruct ancestral relationships, and study the origins of biological diversity.
7. **Physiology:** Physiology is the branch of biology that deals with the study of the functions and mechanisms of living organisms. Physiologists explore how living organisms, from single cells to complex multicellular organisms, carry out vital processes such as metabolism, growth, reproduction, and response to stimuli. They investigate the functioning of organs, tissues, and systems within the body.
8. **Biochemistry:** Biochemistry focuses on the chemical processes and substances that occur within living organisms. Biochemists examine the structure and function of biomolecules such as proteins, carbohydrates, lipids, and nucleic acids. They investigate the metabolic pathways that drive cellular processes, enzymatic reactions, and the molecular basis of genetic information.
9. **Biotechnology:** Biotechnology involves the use of living organisms, or their products, to develop technological applications that benefit humans and the environment. Biotechnologists apply principles from various fields, including genetics, molecular biology, and biochemistry, to develop new drugs, genetically modified crops, biofuels, and diagnostic tools.
10. **Anatomy:** Anatomy is the study of the structure and organization of living organisms. Anatomists investigate the relationships between different body parts, tissues, and organs,

and how they function together as integrated systems. They use techniques such as dissection, microscopy, and medical imaging to explore the intricacies of anatomical structures.

11. **Immunology:** Immunology is the branch of biology that focuses on the study of the immune system and its role in protecting the body against diseases. Immunologists investigate the structure and function of the immune system, the mechanisms of immune response, and the development of vaccines and immunotherapies.

12. **Paleontology:** Paleontology involves the study of fossils and ancient life forms to understand the history of life on Earth. Paleontologists analyze the fossil record to reconstruct past environments, trace the evolution of organisms, and uncover critical events such as mass extinctions.

13. **Neurobiology:** Neurobiology, also known as neuroscience, is the branch of biology that investigates the structure and function of the nervous system. Neurobiologists explore the intricacies of the brain, spinal cord, and peripheral nervous system, studying topics such as neural development, sensory perception, cognition, and behavior.

14. **Biogeography:** Biogeography examines the distribution patterns of organisms across space and time. Biogeographers study the factors that influence the distribution of species, the formation of biomes, and the colonization of new habitats. They contribute to our understanding of the processes that shape biodiversity and the impacts of environmental changes.

15. **Ethology:** Ethology is the scientific study of animal behavior. Ethologists investigate how animals behave in their natural environments, including social interactions, communication, mating rituals, and foraging strategies. They employ observation, experimentation, and theoretical modeling to gain insights into the evolutionary and ecological significance of behavior.

16. **Virology:** Virology is the branch of biology that focuses on the study of viruses. Virologists examine the structure, replication, and evolution of viruses, as well as their interactions with host organisms. They play a crucial role in understanding viral diseases, developing antiviral therapies, and vaccine production.
17. **Marine Biology:** Marine biology is the study of marine organisms and their ecosystems. Marine biologists explore the diverse life forms inhabiting oceans, coral reefs, estuaries, and other aquatic environments. They investigate marine biodiversity, ecological interactions, conservation strategies, and the impacts of human activities on marine ecosystems.
18. **Biophysics:** Biophysics applies the principles and methods of physics to understand biological systems. Biophysicists study biological processes using quantitative techniques, such as mathematical modeling, imaging, and spectroscopy. They explore topics such as molecular interactions, biomechanics, and the physical properties of biological molecules.

These branches of biology represent a broad spectrum of scientific inquiry and provide a deeper understanding of the complexity and interconnectedness of life. As the field of biology continues to advance, interdisciplinary approaches and emerging branches will further expand our knowledge of the natural world.

■ Scientific Method In Biology

The scientific method is a systematic approach used by scientists to investigate and understand the natural world. It provides a framework for conducting experiments, making observations, forming hypotheses, testing predictions, and drawing conclusions based on empirical evidence. In the field of biology, the scientific method plays a crucial role in advancing our knowledge of living organisms and their interactions with the environment.

a. **Observation:** The scientific method begins with careful observation of a phenomenon or a question about the natural world. Scientists keenly observe and gather relevant information to identify patterns, trends, or potential problems that require further investigation. Observations can be made through direct observations of living organisms, experiments, data analysis, or by reviewing existing scientific literature.
b. **Question:** Based on observations, scientists develop specific questions that need to be addressed. These questions are often focused on understanding the mechanisms, processes, or relationships that underlie the observed phenomena. Well-formulated questions guide the entire scientific inquiry process and help scientists formulate hypotheses.
c. **Hypothesis:** A hypothesis is a testable explanation or a proposed answer to the research question. It is based on prior knowledge, observations, and logical reasoning. Hypotheses must be specific, measurable, and falsifiable, allowing them to be tested through experiments or other scientific methods. In biology, hypotheses often involve cause-and-effect relationships, biological mechanisms, or predictions about the behavior of living organisms.
d. **Prediction:** Once a hypothesis is formulated, scientists make predictions that can be tested through experimentation or further observations. Predictions are specific statements that describe the expected outcomes or results if the hypothesis is correct. These predictions serve as benchmarks for evaluating the validity of the hypothesis.
e. **Experimentation:** Experiments are designed to test the predictions derived from the hypothesis. Scientists carefully plan and conduct experiments, controlling variables and manipulating factors to observe the effects on the system

under study. Experimental designs often include a control group, which serves as a baseline for comparison, and one or more experimental groups where specific variables are changed or manipulated.
f. **Data Collection:** During experiments, scientists collect data through careful measurements, observations, or recordings. Data can be quantitative (numerical) or qualitative (descriptive). It is important to collect data systematically and accurately to ensure the reliability and validity of the results. Various tools and techniques, such as microscopes, sensors, data loggers, and statistical analysis, are employed to gather and analyze the data.
g. **Data Analysis:** Once the data is collected, it is analyzed to identify patterns, trends, and relationships. Statistical methods and data visualization techniques are often used to interpret the data and determine if it supports or refutes the hypothesis. Data analysis helps scientists draw meaningful conclusions and make logical inferences based on the evidence gathered.
h. **Conclusion:** Based on the data analysis, scientists draw conclusions regarding the hypothesis. The conclusions may either support or reject the initial hypothesis. It is important to critically evaluate the results, considering any limitations or sources of error that may have influenced the outcomes. If the hypothesis is supported, the conclusions may lead to new questions or further investigations. If the hypothesis is rejected, alternative explanations may be proposed and tested.
i. **Communication:** Scientific findings and conclusions are shared with the scientific community and the public through scientific publications, presentations, conferences, or other forms of communication. This dissemination of knowledge allows for scrutiny, replication, and collaboration among scientists, promoting the advancement of scientific understanding in biology.

j. **Further Exploration:** The scientific method is an iterative process, and the results of one study often lead to new questions and avenues of exploration. Scientists build upon existing knowledge to generate new hypotheses and conduct further research, creating a cycle of continuous discovery and understanding in the field of biology.

■ Tools And Techniques In Biology

In the field of biology, a wide range of tools and techniques are utilized to study and investigate various aspects of living organisms, from the molecular level to ecosystems. These tools and techniques enable scientists to explore the intricate mechanisms of life, understand biological processes, and make significant contributions to the advancement of knowledge in the field. Each tool and technique plays a unique role in unraveling the mysteries of life and advancing our understanding of the living world. The continued development and application of these tools and techniques contribute significantly to the progress of biological research and its various interdisciplinary branches. Let us take a look at some of the essential tools and techniques employed in biology.

1. **Microscopy:** Microscopy is a fundamental tool in biology that allows scientists to observe and study organisms, cells, tissues, and subcellular structures. Different types of microscopes are utilized, including light microscopes, electron microscopes, and fluorescence microscopes. Light microscopes use visible light to magnify and visualize samples, while electron microscopes employ electron beams to achieve higher magnification and resolution, enabling the examination of ultrastructural details. Fluorescence microscopy utilizes fluorescent dyes or tags to visualize specific molecules or structures within cells or tissues.

2. **DNA Sequencing:** DNA sequencing techniques have revolutionized the field of genetics and genomics. These techniques enable the determination of the precise order of nucleotides in DNA molecules. Different methods, such as Sanger sequencing and next-generation sequencing (NGS), are employed for DNA sequencing. Sanger sequencing, based on the chain termination method, was the first widely used technique. NGS techniques, including Illumina sequencing and Oxford Nanopore sequencing, allow for rapid and cost-effective sequencing of entire genomes, facilitating various applications such as genome mapping, identification of genetic variations, and gene expression analysis.
3. **Polymerase Chain Reaction (PCR):** PCR is a powerful molecular biology technique used to amplify specific regions of DNA. It allows researchers to generate a large number of copies of a DNA segment, even from a minuscule starting amount. PCR has applications in various areas, including genetic research, diagnostics, forensics, and biotechnology. Real-time PCR (qPCR) enables quantitative analysis of DNA or RNA molecules, providing valuable insights into gene expression levels and viral load measurements.
4. **Gel Electrophoresis:** Gel electrophoresis is a technique employed to separate and analyze DNA, RNA, or proteins based on their size and charge. It involves the migration of charged molecules through a gel matrix under the influence of an electric field. Agarose gel electrophoresis is commonly used for separating DNA fragments, while polyacrylamide gel electrophoresis is employed for higher resolution analysis of small DNA fragments or proteins. The separated molecules can be visualized by staining with dyes or using specialized detection methods.
5. **Spectroscopy:** Spectroscopy techniques play a crucial role in the analysis of biological molecules and processes. UV-visible

spectroscopy utilizes the absorption of ultraviolet or visible light by molecules to determine their concentrations or characterize their properties. Infrared (IR) spectroscopy analyzes the absorption, transmission, and reflection of infrared light, providing information about molecular structures and functional groups. Nuclear Magnetic Resonance (NMR) spectroscopy and Mass Spectrometry (MS) are advanced techniques used to analyze the structure and composition of biomolecules, including proteins, nucleic acids, and metabolites.

6. **Chromatography:** Chromatography techniques are employed for the separation, purification, and analysis of complex mixtures of molecules. Liquid Chromatography (LC), Gas Chromatography (GC), and High-Performance Liquid Chromatography (HPLC) are commonly used in biological research. LC and HPLC are employed for separating and analyzing compounds in liquid samples, while GC is used for volatile compounds in gaseous samples. Chromatography techniques find applications in the analysis of proteins, nucleic acids, carbohydrates, lipids, and various small molecules.

7. **Cell Culture and Tissue Culture:** Cell culture and tissue culture techniques involve the in vitro growth and maintenance of cells or tissues under controlled conditions. These techniques provide a controlled environment for studying cell behavior, cellular interactions, and the effects of external factors on cells. Cell culture and tissue culture have widespread applications in areas such as drug discovery, regenerative medicine, and the production of biopharmaceuticals.

8. **Field Sampling and Surveys:** In biology, field sampling and surveys are essential for studying organisms in their natural habitats and assessing ecological parameters. Various techniques, including trapping, netting, and mark-and-

recapture methods, are employed to capture and study organisms in their native environments. Field surveys involve collecting data on species diversity, population dynamics, behavior, and interactions within ecosystems, contributing to ecological research and conservation efforts.
9. **Bioinformatics:** Bioinformatics combines biology, computer science, and statistics to manage and analyze biological data, particularly in genomics and proteomics. Bioinformatics tools and techniques enable the storage, retrieval, and analysis of large-scale biological data, facilitating genome sequencing, gene annotation, protein structure prediction, and the identification of genetic variations or disease markers. Bioinformatics plays a vital role in understanding complex biological systems and conducting comparative genomics studies.
10. **Statistical Analysis:** Statistical analysis is an integral part of biological research, allowing scientists to draw meaningful conclusions from experimental data. Statistical methods are used to analyze data, determine the significance of results, and establish correlations or relationships between variables. Statistical techniques help in experimental design, hypothesis testing, data interpretation, and ensuring the validity and reliability of scientific findings.

▪ Careers In Biology

Biology opens up a wide range of exciting and rewarding career opportunities. Whether you have a passion for research, healthcare, conservation, or education, a biology background can pave the way for diverse and impactful careers.
- ✓ **Research Scientist:** Research scientists play a vital role in advancing our understanding of the natural world. They conduct experiments, collect data, and analyze results to

contribute to scientific knowledge. Research scientists can specialize in various areas, such as molecular biology, genetics, ecology, microbiology, or biochemistry. They may work in academic institutions, government agencies, private research firms, or pharmaceutical companies.
- ✓ **Healthcare Professional:** Biology forms the foundation for numerous healthcare careers. Medical doctors (physicians) diagnose and treat illnesses, while surgeons perform surgical procedures. Other healthcare professionals include nurses, pharmacists, physical therapists, and medical laboratory technicians who work in hospitals, clinics, and research institutions. Biological knowledge is crucial for understanding diseases, developing treatments, and promoting overall health and well-being.
- ✓ **Biotechnologist:** Biotechnology utilizes living organisms and their components to develop new products and technologies. Biotechnologists apply biological principles to areas such as pharmaceuticals, agriculture, environmental conservation, and genetic engineering. They may work on developing new drugs, improving crop yields, designing biofuels, or creating genetically modified organisms (GMOs) with beneficial traits.
- ✓ **Environmental Scientist:** Environmental scientists study the natural environment and its interaction with human activities. They assess and mitigate environmental impacts, conduct research on ecosystems, monitor pollution levels, and develop conservation strategies. Environmental scientists often collaborate with policymakers, businesses, and communities to promote sustainable practices and protect biodiversity.
- ✓ **Conservation Biologist:** Conservation biologists work to preserve and restore biodiversity and natural habitats. They study endangered species, develop conservation plans, and advocate for sustainable resource management. Conservation biologists may work in national parks, wildlife reserves, non-

profit organizations, or governmental agencies dedicated to protecting the environment.
- ✓ **Forensic Scientist:** Forensic scientists apply biological knowledge and techniques to assist in criminal investigations. They analyze biological evidence, such as DNA, fingerprints, and bodily fluids, to provide scientific evidence for legal proceedings. Forensic biologists work closely with law enforcement agencies, forensic laboratories, and legal professionals.
- ✓ **Science Writer/Journalist:** Science writers and journalists bridge the gap between scientific research and the general public. They communicate complex scientific concepts in a clear and engaging manner through articles, books, blogs, and media platforms. Science communicators often work for science magazines, news outlets, museums, or educational organizations.
- ✓ **Teacher/Educator:** Biology educators inspire and educate the next generation of scientists. They teach biology at various educational levels, from secondary schools to universities. Biology teachers foster scientific curiosity, conduct laboratory experiments, and help students develop critical thinking and analytical skills.
- ✓ **Wildlife Biologist:** Wildlife biologists study animal behavior, populations, and habitats. They work in natural environments, conducting field research, tracking animal movements, and assessing the impact of human activities on wildlife. Wildlife biologists may collaborate with conservation organizations, national parks, or research institutions.
- ✓ **Pharmaceutical Researcher:** Pharmaceutical researchers focus on developing new drugs and treatments for various diseases. They conduct laboratory experiments, analyze data, and collaborate with medical professionals to bring novel therapies to the market. Pharmaceutical researchers work in

Introduction To Biology

academic institutions, pharmaceutical companies, or research laboratories.

✓ **Genetic Counselor:** Genetic counselors provide guidance and support to individuals and families regarding genetic disorders and inherited conditions. They help people understand genetic risks, interpret test results, and make informed decisions about family planning and healthcare. Genetic counselors work in hospitals, genetic testing centers, and specialized clinics.

✓ **Marine Biologist:** Marine biologists study marine organisms and ecosystems, exploring the diverse life forms found in oceans, seas, and estuaries. They conduct research on marine biodiversity, conservation, and the impacts of climate change. Marine biologists work in research institutions, environmental organizations, or government agencies.

✓ **Botanist:** Botanists specialize in the study of plants, including their classification, growth patterns, and ecological roles. They may research plant genetics, study plant diseases, or work on plant conservation and restoration projects. Botanists may find employment in botanical gardens, agricultural organizations, or research institutions.

✓ **Agricultural Scientist:** Agricultural scientists apply biological knowledge to enhance agricultural practices, improve crop yields, and develop sustainable farming methods. They may focus on plant genetics, soil science, pest management, or livestock production. Agricultural scientists work in research institutions, governmental agencies, or agricultural companies.

✓ **Bioinformatician:** Bioinformatics combines biology, computer science, and statistics to analyze and interpret biological data. Bioinformaticians develop algorithms and computational tools to study DNA sequences, protein structures, and biological networks. They may work in research institutions, pharmaceutical companies, or bioinformatics firms.

- ✓ **Science Policy Analyst:** Science policy analysts bridge the gap between science and policymaking. They evaluate scientific research, analyze its implications, and contribute to the development of science-based policies. Science policy analysts often work for government agencies, non-profit organizations, or research institutions.

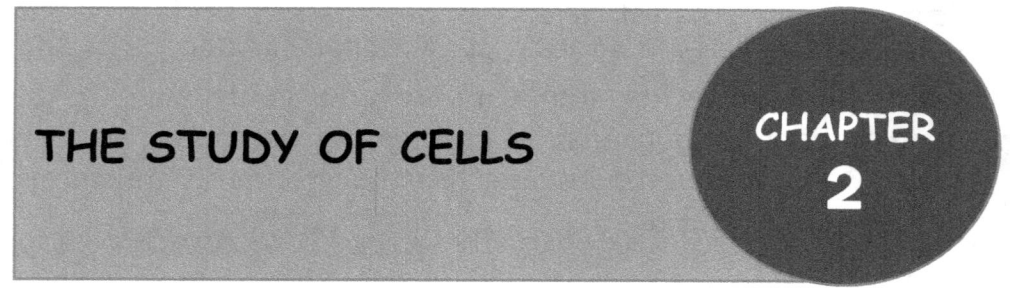

THE STUDY OF CELLS
CHAPTER 2

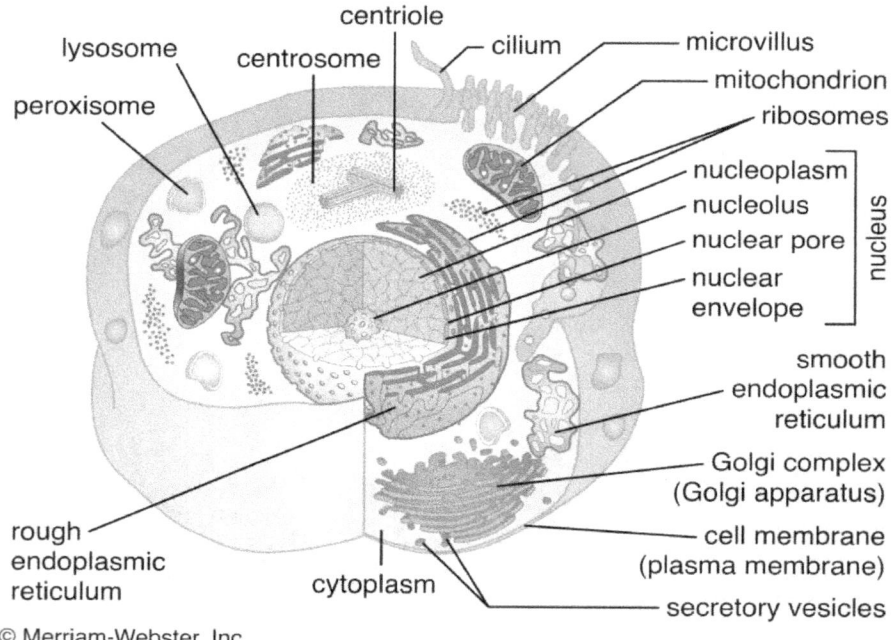

© Merriam-Webster, Inc.

■ Cell Theory

Cell theory is one of the fundamental principles in biology that describes the basic unit of life and the organization of all living organisms. It revolutionized our understanding of life and forms the foundation of modern biology. The cell theory states that:

1. **All living organisms are composed of cells:** Cells are the building blocks of life. Whether organisms are single-celled or

multicellular, they are composed of cells. From microscopic bacteria to complex organisms like humans, every living entity is made up of cells.
2. **The cell is the basic unit of structure and function:** The cell is the smallest functional and structural unit of an organism. Each cell is a self-contained entity that carries out the essential processes of life, including metabolism, reproduction, and response to stimuli. Cells vary in size, shape, and function, but they all share fundamental characteristics.
3. **All cells arise from pre-existing cells:** Cells do not spontaneously generate but arise from pre-existing cells through cell division. This concept, known as cell lineage, is a fundamental principle of cell theory. Every cell is derived from another cell, either through asexual reproduction (such as binary fission or mitosis) or sexual reproduction (where gametes fuse to form a zygote).

■ Historical Development Of Cell Theory

The formulation of cell theory was a collaborative effort built upon the contributions of several scientists over time. Here are the key milestones in the development of cell theory:

i. **Robert Hooke (1665):** Using a simple microscope, Hooke observed cork slices and coined the term "cell" to describe the box-like structures he observed. While these structures were dead plant cells, this marked the first recorded use of the term "cell."

ii. **Anton van Leeuwenhoek (late 17th century):** Leeuwenhoek, a pioneer of microscopy, improved the microscope and made detailed observations of living microorganisms, including bacteria and protists. His observations provided evidence for the existence of single-celled organisms.

iii. **Matthias Schleiden (1838):** Schleiden, a botanist, proposed that plants are composed of cells and that cells are the basic building blocks of plants. This laid the foundation for the plant cell theory.
iv. **Theodor Schwann (1839):** Schwann, a zoologist, extended Schleiden's ideas to animals and proposed that animals are also composed of cells. He formulated the animal cell theory.
v. **Rudolf Virchow (1855):** Virchow, a physician and pathologist, emphasized that cells can only arise from pre-existing cells. He consolidated the principles of cell theory and emphasized the significance of cell division in the growth and development of organisms.

■ Cell Structure And Function

Prokaryotic Cells

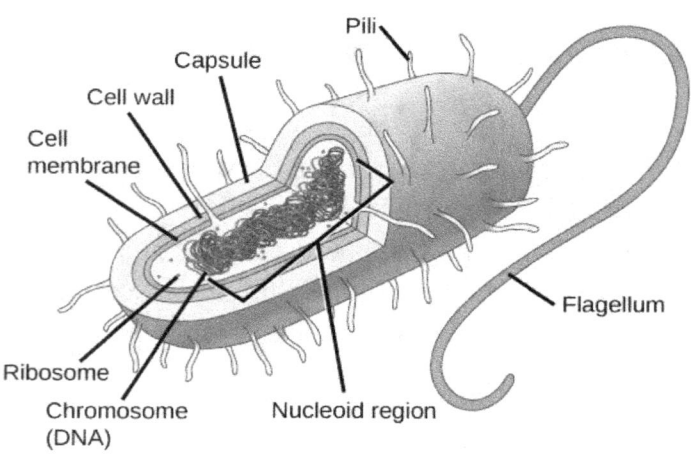

Prokaryotic cells are a type of cell lacking a nucleus and other membrane-bound organelles. They are found in organisms belonging to the domains Bacteria and Archaea. Prokaryotes are single-celled organisms, although some can form multicellular structures like biofilms. In this detailed explanation, we will explore the structure and characteristics of prokaryotic cells.

Cell Envelope

The cell envelope of prokaryotic cells consists of three main components: the cell membrane, cell wall, and in some cases, an outer capsule or slime layer.

a. **Cell Membrane:** The cell membrane, also known as the plasma membrane, is a thin and flexible barrier that encloses the cytoplasm of the cell. It consists of a phospholipid bilayer embedded with proteins. The cell membrane regulates the passage of substances in and out of the cell, maintains cell shape, and is involved in various cellular processes, including energy production and signal transduction.

b. **Cell Wall:** Prokaryotic cells have a cell wall outside the cell membrane, which provides structural support and protection. The composition of the cell wall varies among different groups of prokaryotes. Bacteria typically have a peptidoglycan cell wall, which is a complex mesh-like structure composed of sugars and amino acids. Archaea have cell walls made of different materials, such as pseudopeptidoglycan or other unique molecules.

c. **Capsule/Slime Layer:** Some prokaryotes possess an additional outer layer called a capsule or slime layer. This layer is composed of polysaccharides or proteins and helps protect the cell against desiccation, immune responses, and other environmental challenges. It also aids in cell attachment to surfaces and the formation of biofilms.

Cytoplasm

The cytoplasm is a gel-like substance that fills the interior of the prokaryotic cell. It contains various structures, including the nucleoid region, ribosomes, inclusion bodies, and the cytoskeleton.

a. **Nucleoid:** The nucleoid region is the central region of the cytoplasm where the genetic material of the cell, usually a

circular DNA molecule, is located. Unlike eukaryotic cells, prokaryotes do not have a membrane-bound nucleus. The nucleoid region contains the genetic information necessary for cell replication and function.
b. **Ribosomes:** Ribosomes are responsible for protein synthesis in all cells, including prokaryotes. Prokaryotic ribosomes are smaller in size compared to eukaryotic ribosomes and consist of a small subunit and a large subunit. They float freely in the cytoplasm or may attach to the cell membrane.
c. **Inclusion Bodies:** Prokaryotic cells may contain inclusion bodies, which are storage structures that accumulate various molecules. These can include glycogen granules for energy storage, lipid droplets, sulfur granules, and phosphate granules. Inclusion bodies allow prokaryotes to store essential nutrients for times of scarcity.
d. **Cytoskeleton:** While often overlooked in prokaryotes, recent research has shown the presence of a cytoskeleton-like structure in some species. The prokaryotic cytoskeleton plays a role in maintaining cell shape, cell division, and intracellular organization.

Appendages

Prokaryotic cells can possess various appendages that aid in movement, attachment to surfaces, and the exchange of genetic material.
a. **Flagella:** Flagella are long, whip-like structures used for cell movement. They rotate like a propeller, allowing prokaryotes to move towards or away from stimuli in their environment. The number, arrangement, and structure of flagella can vary among different species.
b. **Pili and Fimbriae:** Pili and fimbriae are thin, hair-like appendages that protrude from the cell surface. Pili are involved in attaching the cell to surfaces or other cells, such as

during the formation of biofilms or the colonization of host tissues. Fimbriae are shorter and more numerous, aiding in attachment to surfaces.
c. **Conjugation Pili:** Conjugation pili, also known as sex pili, are involved in the process of conjugation. Conjugation allows prokaryotes to transfer genetic material, such as plasmids, between cells. This exchange of genetic material contributes to genetic diversity and the spread of advantageous traits.

Metabolic Structures

Prokaryotic cells contain various structures involved in metabolic processes and energy production.

a. **Mesosomes:** Mesosomes are invaginations or extensions of the cell membrane that increase the surface area for cellular respiration and other metabolic reactions. However, recent research suggests that the presence of mesosomes may be an artifact of sample preparation and not a natural feature of prokaryotic cells.
b. **Gas Vesicles:** Some prokaryotes, particularly aquatic bacteria, possess gas vesicles. These vesicles are hollow, gas-filled structures that allow cells to adjust their buoyancy, enabling them to move up or down in water bodies to access optimal oxygen and light conditions.

Prokaryotic cells exhibit remarkable diversity and adaptability. Their simplicity, coupled with their ability to survive in extreme environments and their versatility in metabolic capabilities, has allowed them to thrive in virtually every habitat on Earth. Studying prokaryotic cells provides valuable insights into the fundamental processes of life and helps us understand the intricate interplay between structure and function in living organisms.

Introduction To Biology

Eukaryotic Cells

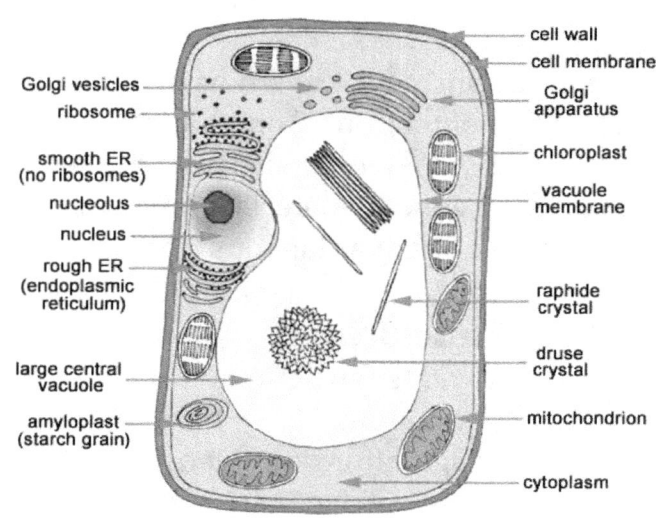

Eukaryotic cells are complex and highly organized cells that make up the majority of organisms on Earth, including animals, plants, fungi, and protists. They are distinguished from prokaryotic cells by the presence of a true nucleus, which houses the cell's genetic material, and membrane-bound organelles that perform specific functions within the cell. Let's explore the various components and functions of eukaryotic cells in detail:

1. **Nucleus:**

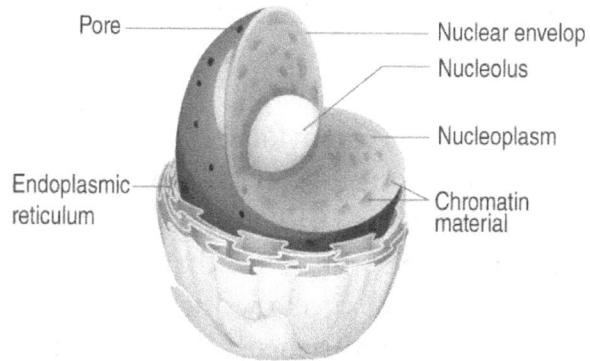

`The nucleus is the most prominent feature of a eukaryotic cell. It is surrounded by a double membrane called the nuclear envelope, which contains nuclear pores that allow the exchange of materials between the nucleus and the cytoplasm.

23

The nucleus contains the cell's DNA in the form of chromosomes, which carry the genetic instructions necessary for cell growth, development, and reproduction. Within the nucleus, a structure called the nucleolus is responsible for the synthesis of ribosomal RNA (rRNA) and the assembly of ribosomes.

2. **Cytoplasm and Cytoskeleton:**

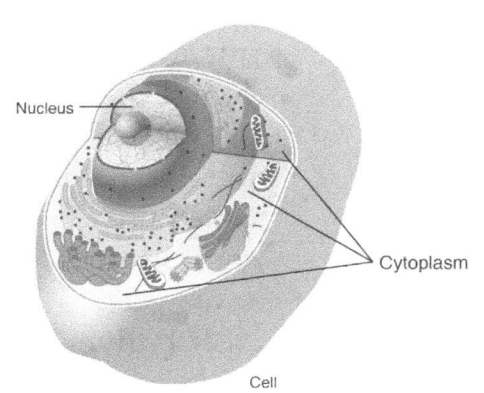

The cytoplasm refers to the gel-like substance that fills the cell between the nucleus and the cell membrane. It consists of a watery solution called the cytosol, which contains various organelles and structures. The cytoskeleton, a network of protein filaments, provides structural support, maintains cell shape, and facilitates cell movement. It is composed of three main components: microfilaments (made of actin), intermediate filaments, and microtubules (made of tubulin).

3. **Endoplasmic Reticulum (ER):**

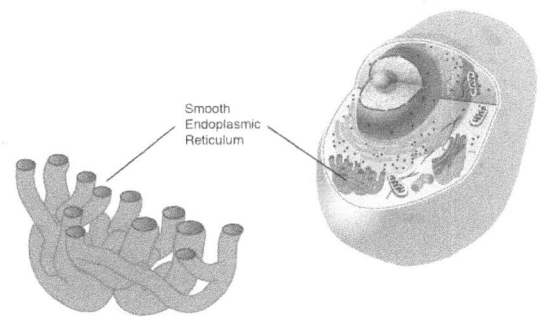

The endoplasmic reticulum is a network of membranous tubules and flattened sacs that extends throughout the cytoplasm. There are two types of ER: rough ER and smooth ER.

Rough ER has ribosomes attached to its surface, giving it a "rough" appearance. It plays a crucial role in protein synthesis, folding, and processing. Smooth ER lacks ribosomes and is involved in lipid metabolism, detoxification of drugs and toxins, and the storage and release of calcium ions.

4. **Golgi Apparatus:**

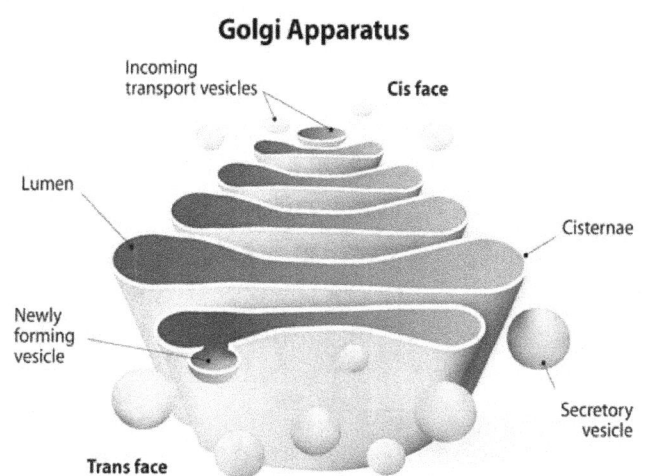

The Golgi apparatus, or Golgi complex, consists of a series of flattened membranous sacs called cisternae. It serves as a major processing and distribution center within the cell. The Golgi apparatus modifies, sorts, and packages proteins and lipids received from the ER for transport to their final destinations, both within the cell and outside of it. It also plays a role in the synthesis of certain carbohydrates.

Introduction To Biology

5. Mitochondria:

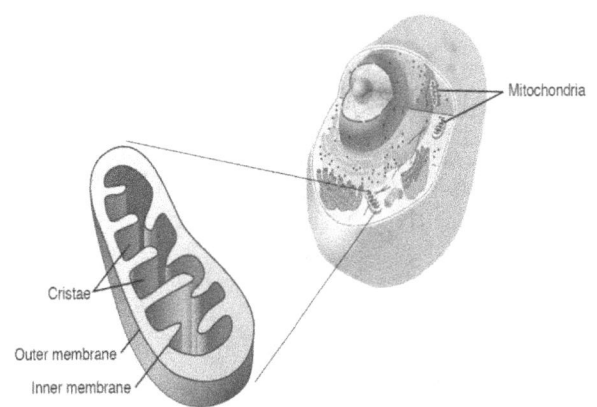

Mitochondria are double-membraned organelles often referred to as the "powerhouses" of the cell. They are responsible for cellular respiration, a process that generates ATP (adenosine triphosphate), the cell's main energy source. Mitochondria have their own DNA and specialized ribosomes, enabling them to produce some of their own proteins. They are highly dynamic organelles that can change their shape and position within the cell based on cellular energy demands.

6. Lysosomes:

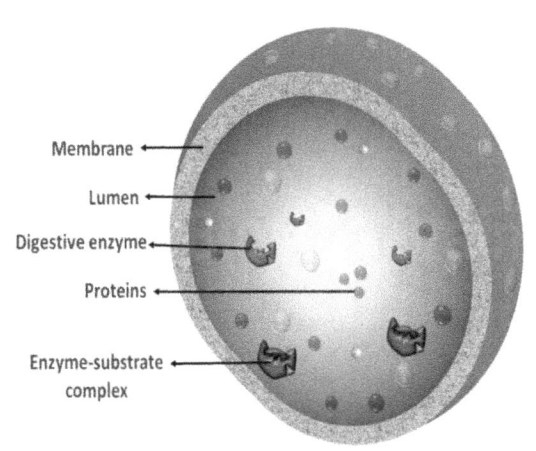

Lysosomes are membrane-bound organelles that contain digestive enzymes called hydrolases. They are involved in the breakdown and recycling of cellular waste materials, damaged organelles, and macromolecules.

Lysosomes are responsible for the degradation of cellular components throu gh a process called autophagy. They also play a role in the immune response by digesting foreign invaders such as bacteria or viruses that enter the cell.

7. Peroxisomes:

Peroxisomes are small, single-membraned organelles involved in various metabolic processes. They contain enzymes called peroxisomal enzymes that participate in the breakdown of fatty acids, the detoxification of harmful substances such as hydrogen peroxide, and the synthesis of certain phospholipids and bile acids. Peroxisomes are especially abundant in liver and kidney cells, where they contribute to detoxification processes.

8. Vacuoles:

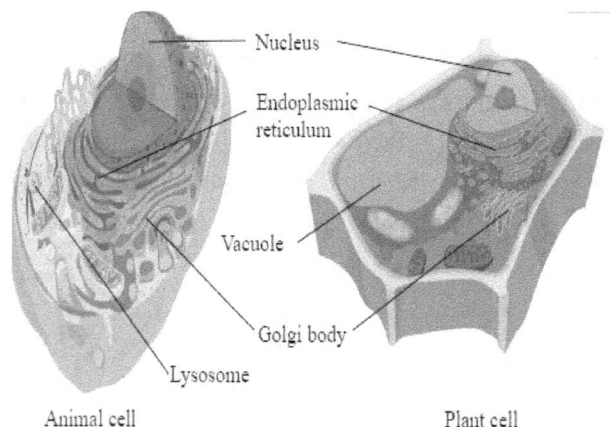

Vacuoles are membrane-bound sacs found primarily in plant cells and some protists. In plant cells, a large central vacuole helps maintain cell turgor pressure and contributes to the regulation of water balance. Vacuoles also serve as storage organelles, storing water, ions, nutrients, pigments, and waste products.

9. Chloroplasts:

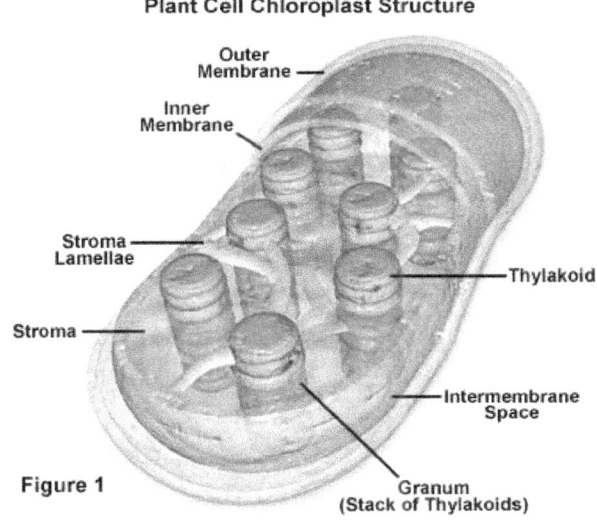

Figure 1 — Plant Cell Chloroplast Structure

Chloroplasts are organelles found in plant cells and some protists. They are responsible for photosynthesis, a process that converts sunlight, carbon dioxide, and water into glucose and oxygen. Chloroplasts contain the pigment chlorophyll, which captures light energy and initiates the synthesis of organic molecules. Like mitochondria, chloroplasts have their own DNA and ribosomes.

10. ll Membrane:

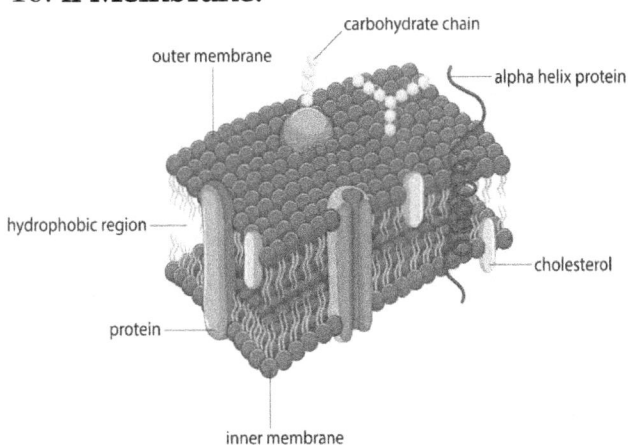

The cell membrane, or plasma membrane, surrounds the eukaryotic cell, separating it from the external environment. It consists of a phospholipid bilayer embedded with proteins. The cell membrane regulates the entry and exit of substances into and out of the cell, facilitates cell-cell communication, and maintains cell shape and integrity. It also plays a role in cell signaling processes and the recognition of self and non-self molecules.

11. Ribosomes:

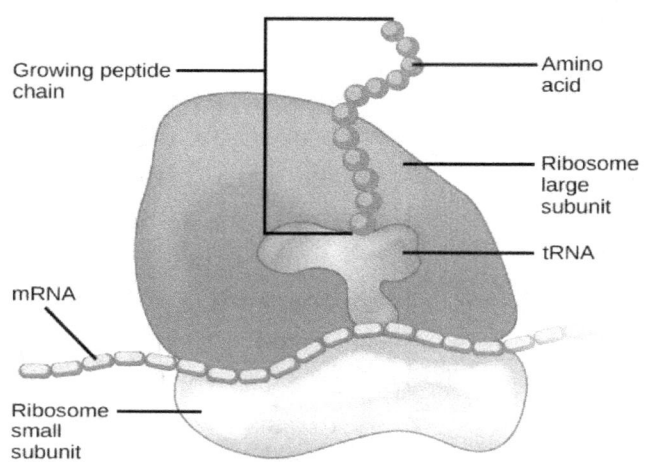

Ribosomes are small, non-membrane-bound organelles responsible for protein synthesis. They can be found free-floating in the cytoplasm or attached to the rough endoplasmic reticulum. Ribosomes read the instructions encoded in messenger RNA (mRNA) and assemble amino acids into proteins through a process called translation.

12. Centrosomes and Centrioles:

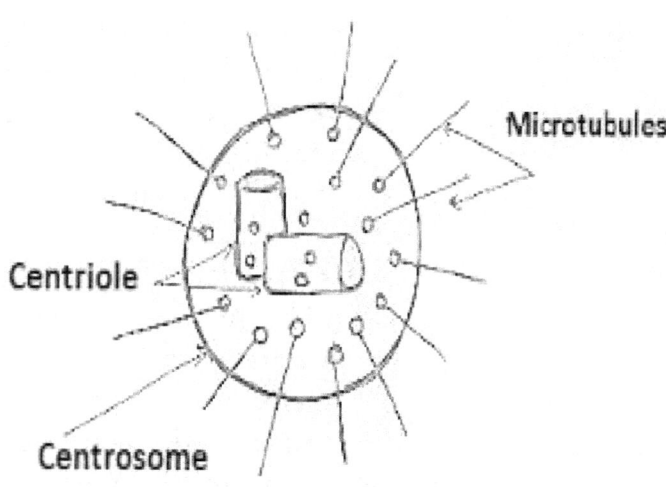

Centrosomes are regions near the nucleus that contain a pair of centrioles. Centrioles are cylindrical structures composed of microtubules and play a crucial role in cell division. They assist in the formation of the mitotic spindle during cell division, which helps separate chromosomes into daughter cells.

13. Cilia and Flagella:

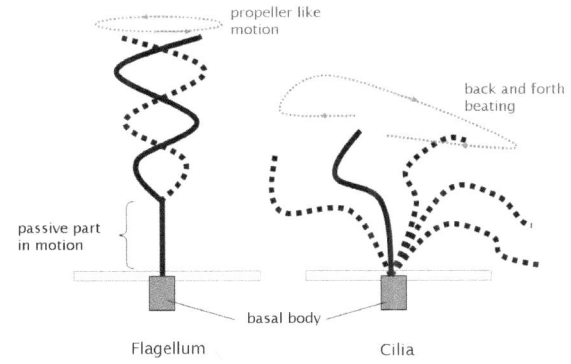

Cilia and flagella are slender, hair-like structures protruding from the cell surface. They are involved in cellular movement. Cilia are shorter and more numerous, while flagella are longer and typically found singly or in pairs. They can be involved in cell locomotion or the movement of fluid and particles along the cell surface.

14. Nucleoplasm and Nuclear Matrix:

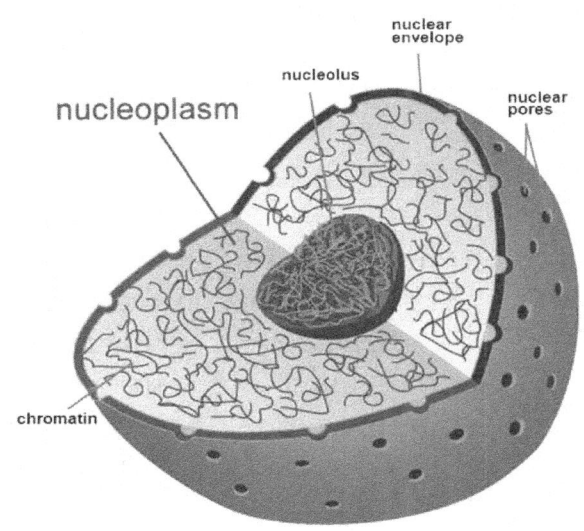

The nucleoplasm is the fluid within the nucleus that surrounds the chromosomes. It contains various proteins, enzymes, and nucleotides necessary for gene expression and DNA replication. The nuclear matrix provides structural support to the nucleus and helps organize the chromosomes.

15. Nuclear Pore Complex:

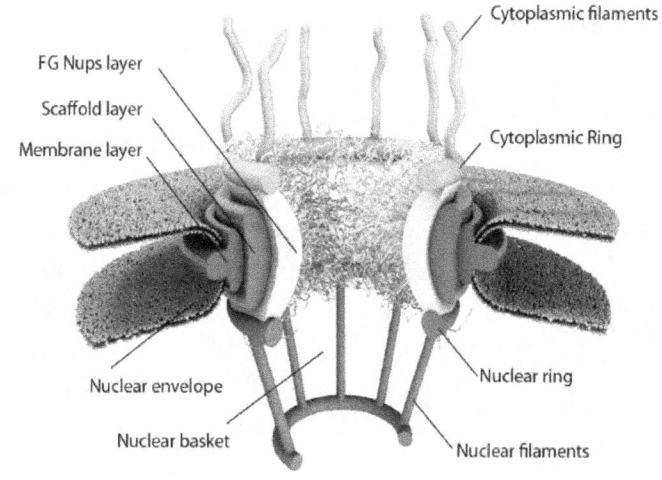

Nuclear pore complexes are protein complexes embedded in the nuclear envelope. They regulate the passage of molecules, such as RNA and proteins, between the nucleus and the cytoplasm. These complexes allow for the exchange of materials required for gene expression, such as the transport of messenger RNA (mRNA) from the nucleus to the cytoplasm for protein synthesis.

16. **Glyoxysomes:** Glyoxysomes are specialized organelles found in plant cells, especially in germinating seeds. They are involved in the conversion of stored lipids into carbohydrates, providing energy for the growing plant embryo until it can perform photosynthesis.

17. **Microvilli:**

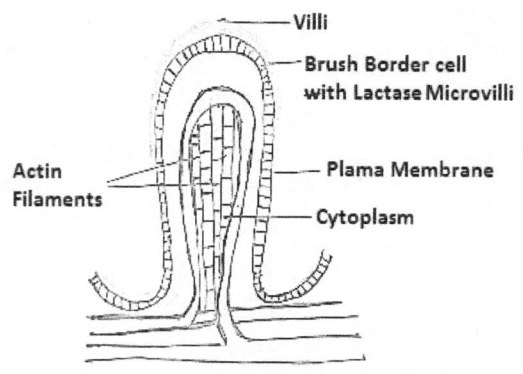

Microvilli are small, finger-like projections on the surface of certain cells, primarily in the lining of the small intestine and kidney tubules. They increase the surface area of the cell, facilitating absorption and secretion processes.

18. **Secretory Vesicles:**

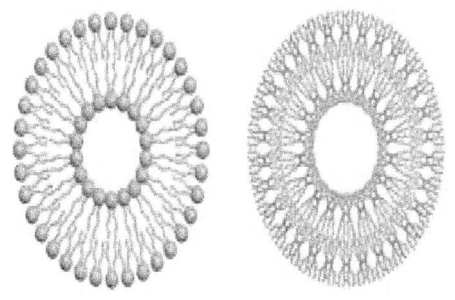

Secretory vesicles are small membrane-bound sacs that transport materials, such as proteins and hormones, from the Golgi apparatus to the cell membrane. When stimulated, these vesicles fuse with the cell membrane, releasing their contents outside the cell through a process called exocytosis.

19. **Contractile Vacuoles:**

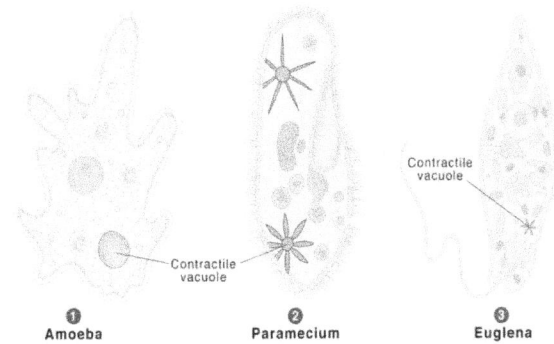

Contractile vacuoles are specialized vacuoles found in certain protists, responsible for regulating water balance and maintaining osmotic pressure. They collect excess water from the cell's cytoplasm and periodically contract, expelling the water outside the cell.

Eukaryotic cells are highly organized and complex structures that contain a nucleus and membrane-bound organelles. These organelles work together to carry out various essential functions, including protein synthesis, energy production, waste removal, and cellular communication. Understanding the structure and function of eukaryotic cells is

Introduction To Biology

vital to comprehending the intricate processes that underlie the functioning of living organisms.

■ Cell Membrane and Transport

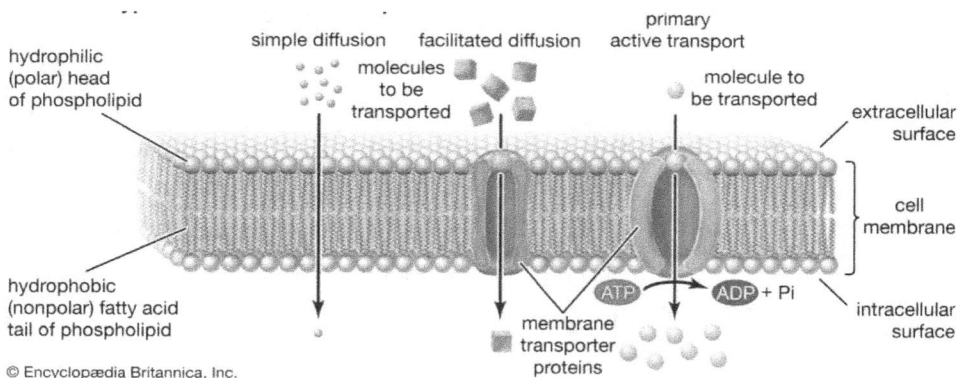

© Encyclopædia Britannica, Inc.

The cell membrane, also known as the plasma membrane, is a vital structure that surrounds the cells of all living organisms. It acts as a selectively permeable barrier, regulating the movement of substances in and out of the cell. The cell membrane is composed of a phospholipid bilayer embedded with proteins, cholesterol, and other molecules. Understanding the structure and functions of the cell membrane, as well as the various mechanisms of cellular transport, is essential to comprehend how cells maintain homeostasis and communicate with their environment.

Structure of the Cell Membrane

The cell membrane is a dynamic and fluid structure that consists of a phospholipid bilayer. Phospholipids are composed of a hydrophilic (water-loving) head and hydrophobic (water-repelling) tails. The hydrophilic heads face the aqueous extracellular and intracellular environments, while the hydrophobic tails form the interior of the membrane. This arrangement creates a selectively permeable barrier that allows the cell to control the movement of substances.

Embedded within the phospholipid bilayer are various proteins that have specific functions. Integral membrane proteins span the entire width of the membrane, while peripheral membrane proteins are attached to either the inner or outer surface. These proteins contribute to cell signaling, transport of molecules, enzymatic activity, and cell adhesion.

Functions of the Cell Membrane

The cell membrane performs several crucial functions:

1. **Selective Permeability:** The cell membrane regulates the passage of substances in and out of the cell. It allows the selective movement of molecules and ions, ensuring the maintenance of appropriate intracellular conditions.
2. **Cell Signaling:** The cell membrane is involved in cell signaling processes, allowing cells to communicate with their environment and other cells. Receptor proteins on the cell surface bind to specific signaling molecules, initiating intracellular responses.
3. **Transport of Molecules:** The cell membrane employs various mechanisms for the transport of molecules across the membrane. These mechanisms include passive transport (diffusion, facilitated diffusion, and osmosis) and active transport (protein pumps, endocytosis, and exocytosis).

Cellular Transport Mechanisms

1. **Diffusion:**

Introduction To Biology

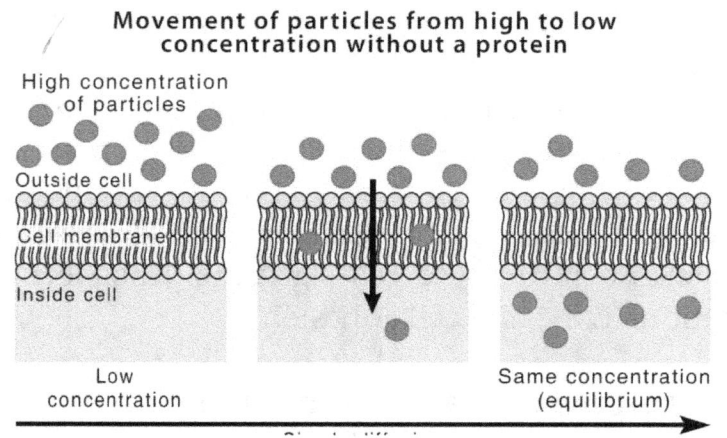

Diffusion is the passive movement of molecules from an area of high concentration to an area of low concentration. Small, non-polar molecules can diffuse directly across the phospholipid bilayer, while larger or charged molecules require specialized protein channels or transporters.

2. Facilitated Diffusion

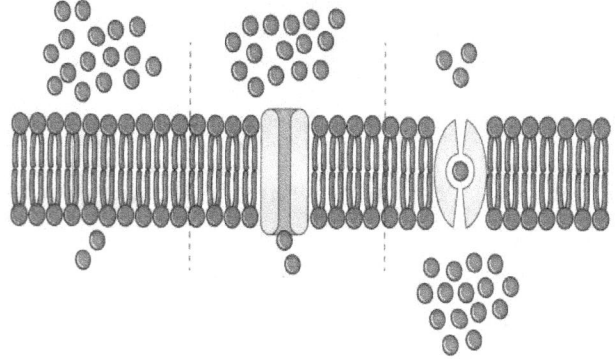

Facilitated diffusion involves the passive transport of molecules across the cell membrane with the help of specific transport proteins. These proteins create channels or carriers that facilitate the movement of molecules down their concentration gradient.

3. Osmosis:

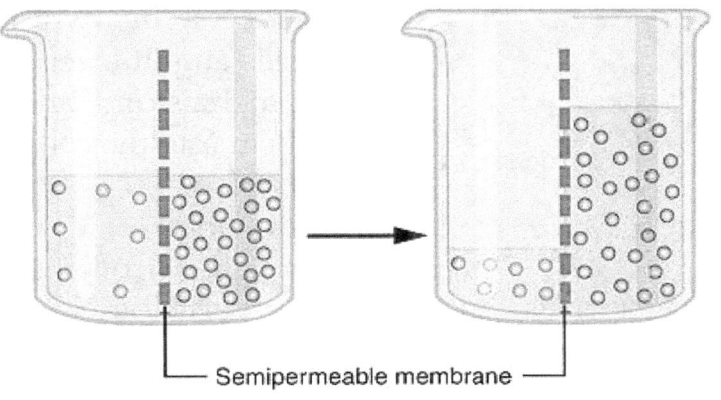

Osmosis is the movement of water molecules across a selectively permeable membrane from an area of lower solute concentration to an area of higher solute concentration. It is essential for maintaining proper water balance in cells.

4. Active Transport:

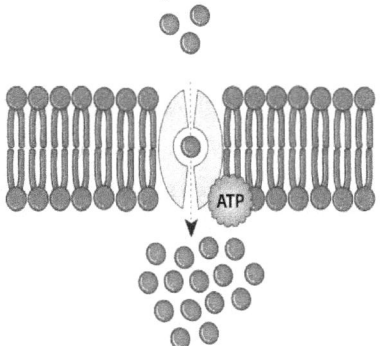

Active transport requires the expenditure of energy (in the form of ATP) to move molecules against their concentration gradient. Protein pumps embedded in the cell membrane utilize energy to transport ions or molecules across the membrane, ensuring their accumulation or removal from the cell.

5. **Endocytosis:**

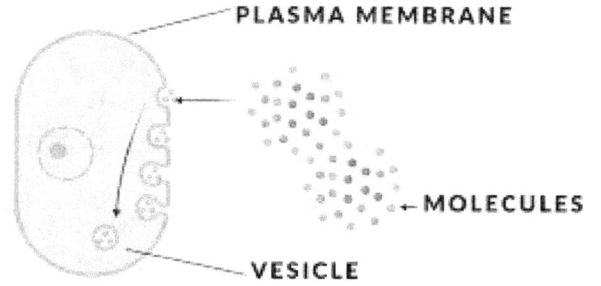

Endocytosis is a process by which cells engulf external materials and bring them into the cell. It involves the formation of vesicles from the cell membrane. Phagocytosis engulfs large particles, while pinocytosis engulfs fluids and dissolved molecules.

6. **Exocytosis:**

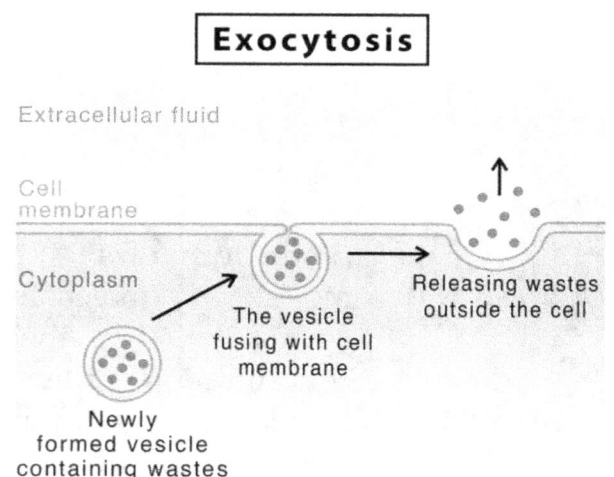

Exocytosis is the process of releasing substances from the cell. Secretory vesicles containing molecules to be expelled fuse with the cell membrane, allowing the contents to be released outside the cell.

Cell Membrane And Homeostasis

The cell membrane plays a crucial role in maintaining cellular homeostasis, which refers to the stable internal environment required for optimal cell function. By controlling the movement of substances, the cell membrane regulates the balance of ions, nutrients, and waste products inside the cell. It also helps

Introduction To Biology

to maintain the proper pH and electrical potential across the membrane, which are essential for cellular processes.

■ Cell Division: Mitosis and Meiosis

Cell division is a fundamental process that allows organisms to grow, develop, and reproduce. There are two main types of cell division: mitosis and meiosis. Let's explore each process in detail:

Mitosis

Mitosis is a type of cell division that occurs in somatic cells, which are non-reproductive cells in multicellular organisms. The primary function of mitosis is the growth, repair, and maintenance of tissues. It consists of four distinct phases: prophase, metaphase, anaphase, and telophase.

Prophase

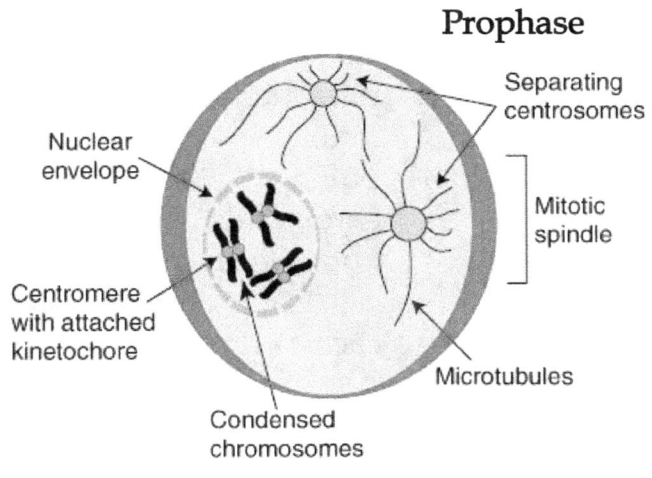

During prophase, the chromatin (loosely packed DNA) in the nucleus condenses and becomes visible as distinct chromosomes. Each chromosome consists of two identical sister chromatids held together by a centromere. The nuclear envelope starts to break down, and the mitotic spindle, composed of microtubules, begins to form.

Metaphase

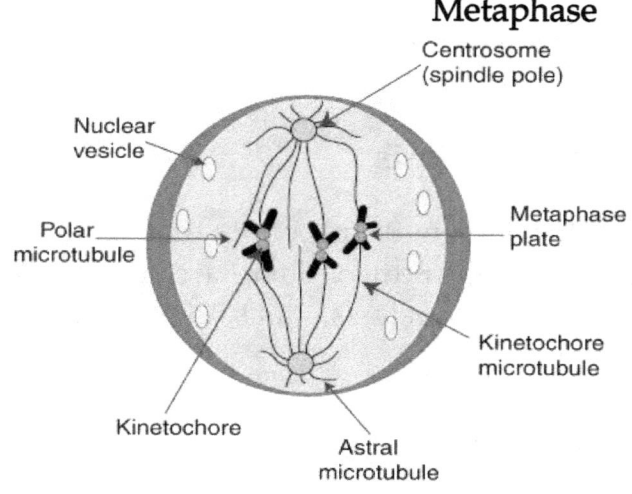

In metaphase, the condensed chromosomes align along the equator of the cell, forming a structure called the metaphase plate. The microtubules of the mitotic spindle attach to the centromeres of the chromosomes, ensuring their proper alignment.

Anaphase

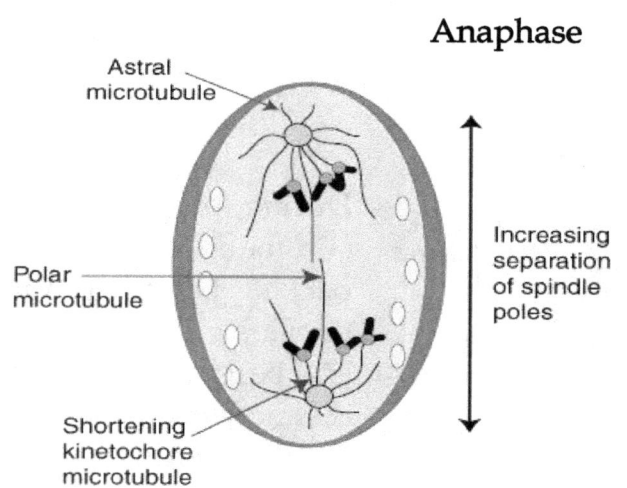

During anaphase, the centromeres split, separating the sister chromatids. The microtubules of the mitotic spindle contract, pulling the sister chromatids towards opposite poles of the cell. This ensures that each daughter cell receives an identical set of chromosomes.

Telophase

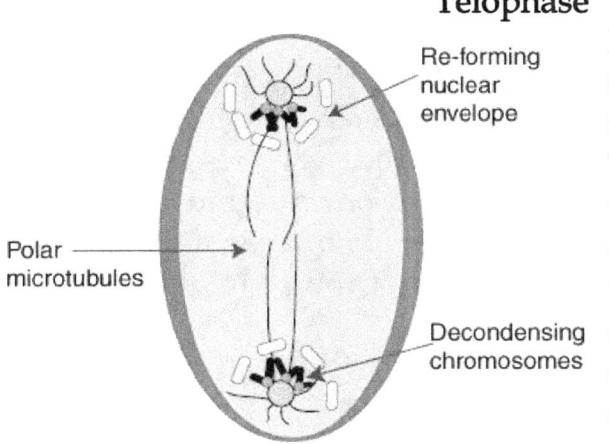

In telophase, the separated chromosomes reach the poles of the cell. The nuclear envelope reforms around each set of chromosomes, creating two new nuclei. The chromosomes begin to decondense, and the mitotic spindle disassembles. Finally, the cell prepares to divide into two daughter cells through cytokinesis.

Cytokinesis

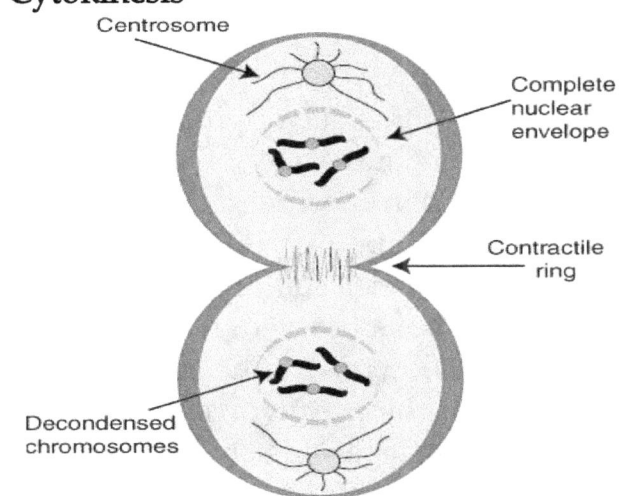

Cytokinesis is the process of physically dividing the cell into two daughter cells. In animal cells, a contractile ring of actin filaments forms at the equator of the cell, constricting the cell membrane and pinching it inward. This forms a cleavage furrow, which deepens until the cell is completely divided into two. In plant cells, a new cell wall, known as the cell plate, forms between the daughter nuclei. The cell plate gradually expands outward until it reaches the cell membrane, separating the cell into two daughter cells.

Meiosis

Meiosis is a specialized type of cell division that occurs in cells involved in sexual reproduction, such as germ cells. The primary function of meiosis is to produce gametes (sperm and eggs) with half the number of chromosomes found in somatic cells. It involves two rounds of division, known as meiosis I and meiosis II, resulting in the production of four haploid daughter cells.

Meiosis I

Prophase I: Prophase I is the longest phase of meiosis and can be further divided into five sub-stages: leptotene, zygotene, pachytene, diplotene, and diakinesis. During prophase I, homologous chromosomes pair up and exchange genetic material through a process called crossing-over. This genetic recombination increases genetic diversity.

Here's a detailed explanation of each sub-stage of prophase I

i. **Leptotene:** Leptotene is the first sub-stage of prophase I. During this stage, the replicated chromosomes condense and become visible under a microscope as thin, thread-like structures. The chromatin fibers further coil and thicken, and individual chromosomes can be observed. The nuclear envelope begins to break down, allowing for the interaction of chromosomes with the cellular machinery involved in meiosis.

ii. **Zygotene:** Zygotene follows leptotene and is characterized by the pairing of homologous chromosomes. The paired chromosomes, called homologous pairs or bivalents, align with each other, beginning at specific sites called the synaptonemal complex. The synaptonemal complex, a protein structure, helps hold the homologous chromosomes together and facilitates the process of crossing-over, which is the exchange of genetic material between homologous chromosomes.

iii. **Pachytene:** Pachytene is the stage during which crossing-over occurs. At this point, the homologous chromosomes fully synapse along their entire length, forming a structure known as a tetrad or bivalent. Crossing-over occurs between non-sister chromatids of homologous chromosomes. Specific sites of contact called chiasmata become visible, where the exchange of genetic material has occurred. This process of genetic recombination increases genetic diversity by creating new combinations of alleles.

iv. **Diplotene:** During diplotene, the homologous chromosomes start to separate slightly but remain connected at the chiasmata. As a result, the tetrad structure becomes more visible. At this stage, the chromosomes continue to condense, and the synaptonemal complex starts to break down. The chromosomes also begin to uncoil slightly, allowing for the resumption of transcriptional activity.

v. **Diakinesis:** Diakinesis is the final sub-stage of prophase I. During this stage, the chromosomes further condense and become highly visible. The nuclear envelope completely breaks down, and the spindle apparatus, composed of microtubules, starts to form. The chiasmata, which represent the sites of crossing-over, move towards the ends of the chromatids, indicating the completion of genetic recombination. The cell is now ready to progress into metaphase I.

Introduction To Biology

STAGES OF PROPHASE OF MEIOSIS I

LEPTOTENE	ZYGOTENE	PACHYTENE	DIPLOTENE	DIAKINESIS
Nuclear membrane	Bivalent forming	Chiasma		Nuclear membrane fragmenting
Replicated chromosomes condense.	Synapsis begins. (Synaptonemal complex forming)	A bivalent has formed and crossing over has occurred.	Synaptonemal complex dissociates.	End of prophase I

Metaphase I

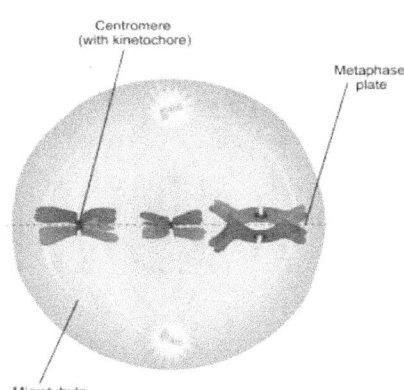

Pairs of homologous chromosomes move to the equator of the cell.

In metaphase I, homologous pairs of chromosomes align along the equator of the cell. Unlike in mitosis, homologous chromosomes, consisting of sister chromatids, line up in pairs rather than individually.

Anaphase I

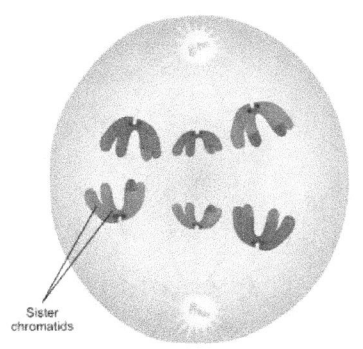

Homologous chromosomes move to the opposite poles of the cell.

During anaphase I, homologous chromosomes separate and move to opposite poles of the cell, pulled by the microtubules of the meiotic spindle. This random separation of chromosomes contributes to genetic variation.

Telophase I

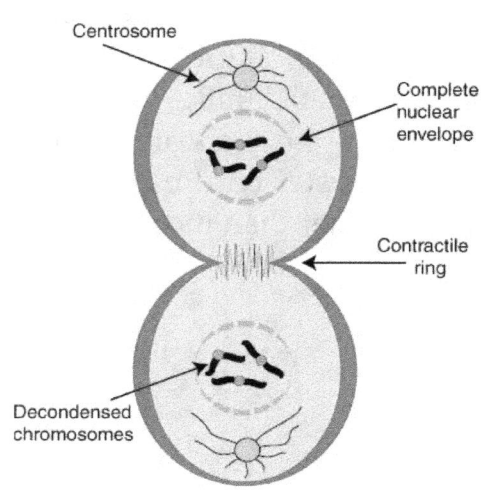

In telophase I, the chromosomes reach the poles of the cell, and the nuclear envelope reforms around each set of chromosomes. Cytokinesis follows, dividing the cell into two daughter cells, each with a haploid set of chromosomes.

Meiosis II

Meiosis II is similar to mitosis and involves the separation of sister chromatids.

Prophase II:

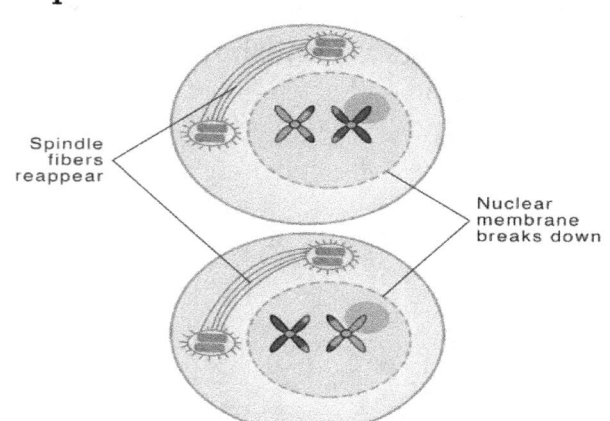

The nuclear envelope breaks down, and the chromosomes condense.

Metaphase II:

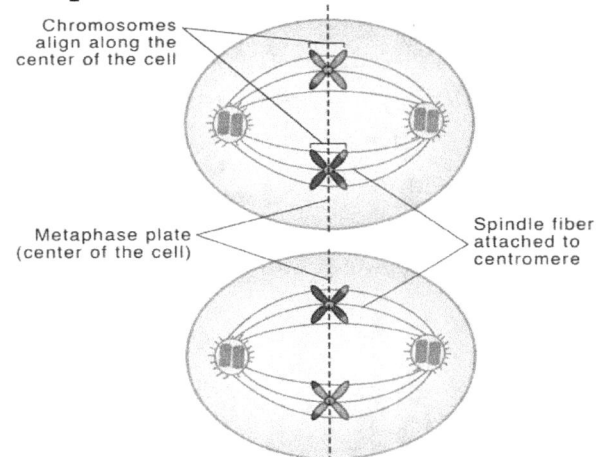

The chromosomes align along the equator of the cell.

Anaphase II:

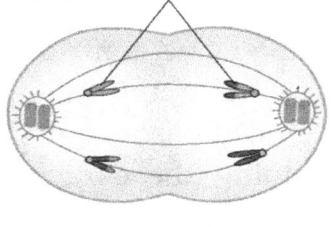
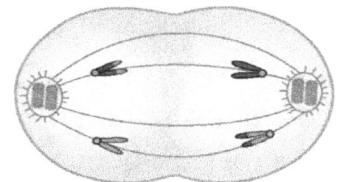

The centromeres split, separating the sister chromatids. The chromatids move to opposite poles of the cell.

Telophase II

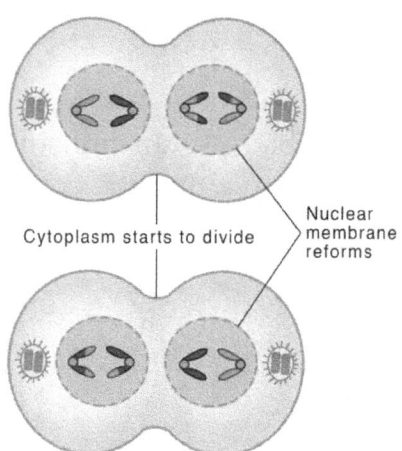

The chromosomes reach the poles of the cell, and the nuclear envelope reforms around each set of chromosomes. Cytokinesis follows, resulting in the formation of four haploid daughter cells.

Significance Of Meiosis

Meiosis is a fundamental process in sexual reproduction that promotes genetic diversity, ensures the maintenance of chromosome number, reduces the genetic load, and contributes to evolutionary processes. Its significance extends to the generation of unique offspring, adaptation to changing environments, and the formation of new species. Understanding the complexities and importance of meiosis is essential for comprehending the diversity and dynamics of life on Earth

1. **Genetic Variation:** Meiosis plays a crucial role in generating genetic diversity. During prophase I of meiosis, homologous chromosomes pair up and undergo crossing-over, where genetic material is exchanged between non-sister chromatids. This recombination of genetic material creates new combinations of alleles, contributing to genetic diversity in offspring. Meiosis also promotes the independent assortment

of chromosomes, resulting in unique combinations of paternal and maternal chromosomes in gametes.
2. **Maintenance of Chromosome Number:** Meiosis ensures the maintenance of the correct chromosome number in sexually reproducing organisms. By halving the chromosome number, meiosis produces haploid gametes (sperm and eggs) with half the number of chromosomes found in somatic cells. When these gametes fuse during fertilization, the resulting zygote restores the diploid chromosome number necessary for normal development.
3. **Reduction of Genetic Load:** Meiosis helps to reduce the accumulation of harmful mutations and genetic abnormalities. During meiosis, homologous chromosomes pair up and undergo a process called synapsis, allowing for the repair of DNA damage. Furthermore, meiotic checkpoints ensure the elimination of cells with chromosomal abnormalities. By removing potentially harmful genetic variations, meiosis promotes the production of healthy offspring.
4. **Evolutionary Advantage:** Meiosis has played a crucial role in evolution by facilitating genetic diversity. The generation of new combinations of alleles through crossing-over and independent assortment provides a substrate for natural selection. This genetic variation allows populations to adapt to changing environments, increasing their chances of survival and reproductive success.
5. **Sexual Reproduction:** Meiosis is intimately linked to sexual reproduction. The production of haploid gametes and their subsequent fusion during fertilization leads to the formation of genetically unique offspring. Sexual reproduction allows for the mixing of genetic material from two parent organisms, increasing the potential for genetic diversity and adaptive traits in populations.

6. **Speciation:** Meiosis contributes to the formation of new species. In sexually reproducing organisms, meiosis facilitates reproductive isolation by creating genetic barriers between populations. Over time, genetic differences accumulate through meiosis and subsequent generations, leading to reproductive incompatibility and the formation of distinct species.

Introduction To Biology

GENETICS AND HEREDITY — CHAPTER 3

Genetics is the scientific study of heredity, which encompasses the processes by which traits are passed from parents to offspring. It investigates the mechanisms and patterns of inheritance, the variation and diversity observed in living organisms, and the molecular basis of genetic information. By studying genetics, scientists seek to uncover the fundamental principles that govern the transmission of genetic traits and how these traits contribute to the overall diversity and functioning of organisms.

Key Concepts in Genetics:

i. **Genes:** Genes are the basic units of heredity. They are segments of DNA that contain the instructions for building and maintaining an organism. Genes determine specific traits, such as eye color, height, and susceptibility to certain diseases. They provide the blueprint for the production of proteins and other molecules that are essential for the structure and function of cells and organisms. Genes are located on chromosomes, which are structures within the nucleus of a cell.

ii. **Alleles:** Alleles are alternative forms or variants of a gene. Each gene may have different alleles, and individuals inherit two copies of each gene, one from each parent. Alleles can be dominant or recessive, with dominant alleles being expressed in the phenotype even if only one copy is present, while recessive alleles are expressed only if two copies are present. Alleles can influence the expression of a particular trait. For

Introduction To Biology

example, the gene responsible for eye color may have different alleles, such as blue, brown, or green.

iii. **Genotype and Phenotype:** Genotype refers to the genetic makeup of an organism, including all the alleles it possesses. Phenotype, on the other hand, refers to the observable physical and biochemical characteristics of an organism, which result from the interaction between genes and the environment. The phenotype can be influenced by both genetic and environmental factors. For example, an individual may have the genotype for tall height, but if they are malnourished, their phenotype may be shorter than expected.

iv. **Mendelian Inheritance:** Gregor Mendel, known as the father of genetics, formulated the basic principles of inheritance through his experiments with pea plants in the 19th century. Mendelian inheritance describes patterns of inheritance governed by the segregation and independent assortment of alleles during gamete formation. Mendel discovered that traits are inherited as discrete units, with each parent contributing one allele for each trait to their offspring. He also observed that the inheritance of different traits is independent of each other, known as the principle of independent assortment.

v. **Chromosomal Basis of Inheritance:** In addition to genes, chromosomes play a crucial role in inheritance. Different genes are located on specific positions of chromosomes, and their inheritance patterns are influenced by the behavior of chromosomes during meiosis, the specialized cell division that produces gametes (sperm and eggs). Chromosomes are structures composed of DNA and proteins that carry genetic information. In humans, for example, individuals typically have 23 pairs of chromosomes, with one set inherited from each parent.

vi. During meiosis, homologous chromosomes pair up and undergo crossing-over, where genetic material is exchanged between non-sister chromatids. This recombination of genetic material creates new combinations of alleles and contributes to genetic diversity. Furthermore, the segregation of chromosomes during meiosis ensures that each gamete receives one copy of each chromosome, resulting in the random assortment of alleles.

vii. **Molecular Genetics:** Molecular genetics focuses on the study of the structure and function of genes at the molecular level. It involves examining DNA, RNA, and proteins to understand how genetic information is stored, replicated, and expressed. Molecular genetics encompasses a range of techniques and methodologies, such as DNA sequencing, PCR (polymerase chain reaction), gene cloning, and gene expression analysis. It has revolutionized the field of genetics and provided insights into the intricate molecular mechanisms underlying inheritance and gene function.

- ## Applications Of Genetics

1. **Medical Genetics:** Understanding the genetic basis of diseases has significant implications for diagnosis, treatment, and prevention. Medical genetics investigates inherited disorders, genetic testing, genetic counseling, and advancements in gene therapy. It allows for the identification of genetic risk factors, the development of personalized medicine approaches, and the understanding of the genetic basis of complex diseases such as cancer, cardiovascular disorders, and neurological conditions.
2. **Evolutionary Genetics:** Genetics plays a vital role in our understanding of evolution and the diversity of life on Earth. By studying genetic variation and evolutionary processes, genetics contributes to our understanding of how species evolve over time and adapt to changing environments. It

helps uncover the genetic mechanisms underlying speciation, the formation of new species, and the processes that lead to the development of diverse traits and adaptations.
3. **Agricultural Genetics:** Genetics has significant applications in agriculture and crop improvement. It plays a crucial role in enhancing crop plants and livestock through selective breeding and genetic engineering. Genetics allows for the identification and manipulation of genes responsible for desirable traits, such as disease resistance, improved yield, and nutritional quality. This knowledge contributes to the development of more productive and resilient agricultural systems.
4. **Forensic Genetics:** DNA profiling and analysis are essential tools in forensic science for identifying individuals, establishing paternity, and solving crimes. Genetic markers, such as short tandem repeats (STRs), are used to compare DNA samples and determine genetic relatedness. Forensic genetics also involves the study of ancient DNA to understand human history and migration patterns.
5. **Conservation Genetics:** Genetics plays a crucial role in the conservation and management of endangered species and biodiversity. It helps assess the genetic diversity within populations, identify individuals for breeding programs, and understand the genetic factors contributing to population declines. Conservation genetics provides insights into the impact of habitat fragmentation, climate change, and human activities on the genetic health and viability of species.

■ Mendelian Inheritance

Mendelian inheritance, also known as Mendelian genetics or classical genetics, refers to the principles of inheritance first elucidated by Gregor Mendel in the mid-19th century. Mendel's groundbreaking experiments with pea plants laid the foundation for understanding how traits are passed from one generation to the next. Mendelian inheritance provides a fundamental framework for understanding how traits are passed from one generation to the next. Through his experiments with pea plants, Gregor Mendel established the laws of segregation and independent assortment, which govern the inheritance of traits. These principles, along with the concepts of dominance and recessiveness, genotype and phenotype, and Punnett squares, have revolutionized our understanding of genetics and form the basis for modern genetic research.

Mendel's Experiments

This Austrian monk and botanist, conducted a series of groundbreaking experiments between 1856 and 1863 using pea plants (Pisum sativum). His experiments laid the foundation for our understanding of the principles of inheritance and established the field of modern genetics. Gregor Mendel, often hailed as the father of modern genetics, conducted groundbreaking experiments in the mid-19th century that laid the foundation for our understanding of inheritance. Mendel's experiments focused on pea plants, particularly the garden pea (Pisum sativum), due to their easy cultivation, distinct traits, and ability to self-pollinate.

Mendel's First Law, also known as the Law of Segregation, emerged from his investigations into single traits. He selected purebred pea plants exhibiting two distinct forms of a trait, such as tall and short. Mendel carefully controlled the pollination process, ensuring the transfer of pollen from the stamen of one plant to the pistil of another. This meticulous cross-breeding allowed him to observe the traits in the next generation.

In Mendel's monohybrid crosses, where he studied a single trait, he found that the first generation (F1) consistently displayed only one of the parental traits, such as all plants being tall. However, when he allowed these F1 plants to self-pollinate, the second generation (F2) showed a 3:1 ratio of dominant to recessive traits. This indicated that the hidden trait in the F1 generation reappeared in the F2 generation, revealing a pattern of inheritance.

Law of Segregation:

The Law of Segregation states that during the formation of gametes (sex cells), the alleles (alternative forms of a gene) segregate from each other, so that each gamete receives only one allele for each trait. This separation occurs because alleles for a trait are located on homologous chromosomes that segregate during meiosis.

Mendel's Second Law, the Law of Independent Assortment, evolved from his dihybrid crosses. In these experiments, Mendel examined the inheritance of two different traits simultaneously, such as seed color and seed shape. He observed that the inheritance of one trait did not influence the inheritance of the other. The traits assorted independently, leading to a variety of combinations in the offspring. Mendel's meticulous record-keeping and statistical analysis were instrumental in revealing the patterns of inheritance. His experiments demonstrated that traits are inherited in discrete units (now known as genes), and their expression is determined by the interaction of dominant and recessive alleles.

Law of Independent Assortment:

The Law of Independent Assortment states that alleles for different traits segregate independently of one another during the formation of gametes. This means that the inheritance of one trait does not influence the inheritance of another trait unless they are located on the same chromosome.

Although Mendel's work initially received little recognition, it gained prominence in the early 20th century when scientists rediscovered his experiments and realized their profound implications for genetics. Mendel's laws became the cornerstones of classical genetics, providing a framework for understanding the hereditary transmission of traits and setting the stage for the modern field of genetics.

- ✓ **Punnett Squares:** *Mendel used Punnett squares as visual tools to predict the possible genotypes and phenotypes of offspring from specific crosses. Punnett squares involve combining the alleles from each parent to determine the genetic combinations that can occur in the offspring.*
- ✓ **Dihybrid Crosses and the Principle of Independent Assortment:** *Mendel expanded his experiments to study the inheritance of two different traits simultaneously, such as seed color and seed shape. These dihybrid crosses involved crossing purebred plants that differed in two traits (e.g., YYRR and yyrr). Mendel discovered that the inheritance of one trait did not influence the inheritance of another. This led to the formulation of the Law of Independent Assortment.*
- ✓ **Statistical Analysis:** *Mendel's meticulous approach to record-keeping and statistical analysis of his experimental results played a crucial role in his discoveries. He compared his observed results to the expected ratios predicted by his laws of inheritance and found them to be remarkably close.*
- ✓ **Publication and Reception:** *Mendel presented his findings in 1865, in a paper titled "Experiments on Plant Hybridization." Unfortunately, his work went largely unnoticed at the time, and it was only in the early 20th century that his experiments gained recognition and became the foundation of modern genetics.*

Dominant and Recessive Traits

Mendel observed that certain traits were dominant over others. Dominant traits are expressed when at least one copy of the dominant allele is present, whereas recessive traits are only expressed when both copies of the gene carry the recessive allele. Dominant alleles are represented by uppercase letters (e.g., "T" for tall), while recessive alleles are represented by lowercase letters (e.g., "t" for short).

Genotype and Phenotype

Mendel introduced the terms genotype and phenotype to describe the genetic and physical characteristics, respectively, of an organism. The genotype refers to the combination of alleles an individual possesses for a particular trait, while the phenotype is the observable expression of the trait. For example, in Mendel's experiments, plants with the genotype "Tt" (heterozygous) had the phenotype of tall plants, as the dominant allele for tallness (T) masked the recessive allele for shortness (t).

Punnett Squares

To predict the possible outcomes of a cross between individuals with known genotypes, Mendel developed Punnett squares. Punnett squares are grids that help determine the probability of different genotypes and phenotypes in the offspring. By combining the alleles from each parent, the Punnett square provides a visual representation of the possible genetic combinations and their corresponding phenotypes.

Mendel conducted monohybrid crosses to study the inheritance of a single trait, while dihybrid crosses involved the simultaneous inheritance of two different traits. In dihybrid crosses, Mendel discovered that alleles for different traits assort independently of one another, in accordance with the Law of Independent Assortment.

Mendelian Inheritance Patterns

Mendelian inheritance follows three main patterns:
a. **Dominant-Recessive Inheritance:** In this pattern, a dominant allele masks the expression of a recessive allele. The dominant allele is inherited in a Mendelian fashion, while the recessive allele is expressed only when both copies are present.
b. **Codominance:** In codominance, both alleles are fully expressed in the heterozygous condition, resulting in a phenotype that displays characteristics of both alleles. For example, in humans, the ABO blood type system demonstrates codominance, where individuals can have blood type A, B, AB, or O.
c. **Incomplete Dominance:** In incomplete dominance, neither allele is fully dominant over the other, resulting in an intermediate phenotype in the heterozygous condition. An example is the inheritance of flower color in snapdragons, where crossing a red-flowered plant with a white-flowered plant produces pink flowers in the offspring.

Mendelian Inheritance in Humans

Mendelian inheritance principles can be applied to the inheritance of traits in humans. Many human traits, such as eye color, blood type, and certain genetic disorders, follow Mendelian patterns of inheritance. However, human inheritance can be more complex due to the involvement of multiple genes and the influence of environmental factors.

Genetic Variation

Genetic variation refers to the diversity of genetic information within a population or species. It is a fundamental concept in genetics and plays a crucial role in evolution, adaptation, and the overall functioning of biological systems. Let's delve into the details of genetic variation and explore its sources, types, and significance.

Sources of Genetic Variation:

Genetic variation arises through several mechanisms:

1. **Mutation:** Mutations are spontaneous changes in the DNA sequence of an organism's genome. They can occur due to errors during DNA replication, exposure to mutagens (such as radiation or chemicals), or as a result of natural processes. Mutations can introduce new genetic variations into a population.
2. **Genetic Recombination:** Genetic recombination occurs during sexual reproduction. It involves the shuffling and exchange of genetic material between homologous chromosomes, leading to new combinations of alleles. This process enhances genetic diversity by creating unique combinations of genetic information.
3. **Gene Flow:** Gene flow refers to the movement of genes between populations. It occurs when individuals migrate and reproduce with individuals from other populations, introducing new genetic material. Gene flow can increase genetic diversity within populations and decrease genetic differences between populations.
4. **Genetic Drift:** Genetic drift refers to the random changes in allele frequencies within a population due to chance events. It has a more significant impact on smaller populations, where random fluctuations can have a greater effect on genetic

diversity. Genetic drift can lead to the loss of rare alleles or the fixation of certain alleles in a population.

Types of Genetic Variation

Genetic variation can manifest in various forms:
1. **Single-Nucleotide Polymorphisms (SNPs):** SNPs are the most common type of genetic variation. They are single-nucleotide differences in the DNA sequence between individuals. SNPs can occur within coding regions of genes or in non-coding regions, and they can influence gene expression, protein structure, and susceptibility to diseases.
2. **Insertions and Deletions (Indels):** Indels are variations in the number of nucleotides within a DNA sequence. They can range from a single base pair insertion or deletion to larger insertions or deletions, such as gene duplications or deletions.
3. **Copy Number Variations (CNVs):** CNVs involve variations in the number of copies of a particular DNA segment. These variations can range from duplications or deletions of entire genes to larger structural changes in the genome.
4. **Structural Variations:** Structural variations encompass larger-scale alterations in the genome, such as inversions, translocations, or chromosomal rearrangements. These variations can have significant effects on gene function and can lead to genetic disorders or adaptations.

Significance of Genetic Variation:

Genetic variation is of utmost importance in several biological processes:
1. **Adaptation and Evolution:** Genetic variation provides the raw material for natural selection and adaptation. It allows

populations to respond to changing environments and increases the likelihood of individuals possessing advantageous traits that enhance survival and reproduction.
2. **Disease Susceptibility:** Genetic variation plays a role in determining an individual's susceptibility to certain diseases. Some genetic variations increase the risk of developing specific disorders, while others confer resistance or protection against certain diseases.
3. **Genetic Diversity and Stability:** High genetic diversity within populations promotes their long-term viability and resilience. It enhances the ability of populations to withstand environmental changes, resist diseases, and maintain overall stability.
4. **Forensic Identification and Parentage Testing:** Genetic variation in the form of DNA markers is used in forensic investigations to establish identity, determine relationships between individuals, and resolve cases of disputed parentage.
5. **Biomedical Research:** Genetic variation provides the basis for studying the genetic basis of diseases, drug responses, and other biological processes. Understanding genetic variation helps in the development of personalized medicine and targeted therapies.

■ DNA and Genes

(DNA and Genes: The Blueprint of Life)

DNA (deoxyribonucleic acid) and genes are fundamental components of living organisms. They hold the instructions necessary for the development, functioning, and inheritance of traits in all forms of life. Understanding the structure and function of DNA, as well as the role of genes, is crucial for unraveling the complexities of genetics and the mechanisms underlying the diversity of life.

Introduction To Biology

DNA and genes are the building blocks of life. DNA carries the genetic information in a unique code, and genes provide the instructions for the synthesis of proteins and other functional molecules. Understanding the structure, replication, and expression of DNA, as well as the role of genes, has profound implications for our understanding of genetics, evolution, and the complex mechanisms underlying life itself. The ongoing advancements in genomics and genetic engineering continue to expand our knowledge and open up new possibilities for medical research, personalized medicine, and the improvement of various aspects of life on Earth.

DNA Structure

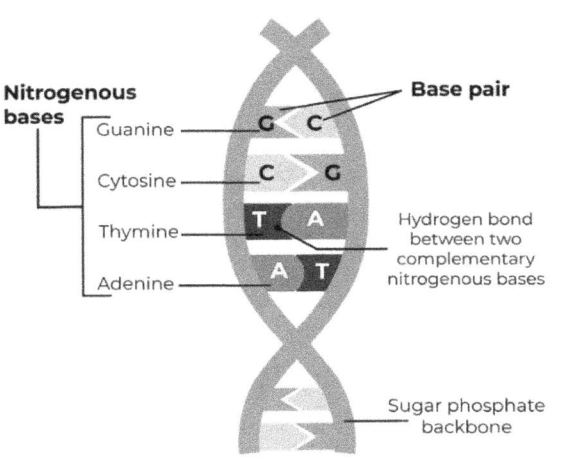

DNA is a double-stranded molecule that forms a unique helical structure known as a double helix. Each strand of the DNA molecule consists of a sugar-phosphate backbone, with nitrogenous bases extending inward and forming hydrogen bonds between the two strands. The four nitrogenous bases are adenine (A), thymine (T), cytosine (C), and guanine (G). Adenine pairs with thymine, and cytosine pairs with guanine, forming complementary base pairs.

Genes

Genes are specific segments of DNA that contain the instructions for producing functional molecules, such as proteins or RNA molecules. Each gene is associated with a particular trait or characteristic. Genes are located on chromosomes, which are thread-like structures within the cell nucleus. Humans have approximately 20,000-25,000 genes.

- **Genetic Code:** The genetic code is the set of rules that specifies how the information in DNA is translated into the production of proteins. The code is based on the sequence of nucleotides (triplets of bases) in DNA, known as codons. Each codon codes for a specific amino acid or serves as a start or stop signal during protein synthesis.
- **Gene Expression:** Gene expression refers to the process by which the information in a gene is used to synthesize a functional product, such as a protein. Gene expression is regulated in a highly complex manner, and it is influenced by various internal and external factors. Different cell types express different sets of genes, leading to the diversity of cell functions and characteristics in multicellular organisms.
- **DNA Mutations:** DNA mutations are changes that occur in the sequence of nucleotides in DNA. Mutations can be caused by various factors, including errors during DNA replication, exposure to certain chemicals or radiation, and environmental factors. Mutations can alter the genetic code, potentially affecting protein structure and function. Some mutations can lead to genetic disorders, while others may have no noticeable effect or even provide an evolutionary advantage.
- **Inheritance:** Genes and DNA are the basis of inheritance. The specific combination of genes inherited from parents determines an individual's traits and characteristics. In sexual reproduction, offspring inherit half of their genetic material from each parent, resulting in genetic variation and the potential for unique combinations of traits.
- **Genomics and Genetic Engineering** Advances in DNA sequencing technologies have revolutionized the field of genomics, which involves studying the complete set of an

organism's genes (genome). Genomics enables researchers to investigate gene functions, understand disease mechanisms, and develop personalized medicine approaches. Genetic engineering techniques, such as gene editing using CRISPR-Cas9, allow for precise modification of DNA, opening up possibilities for medical applications and agricultural advancements.

DNA Replication

One of the remarkable properties of DNA is its ability to replicate itself accurately. During cell division, the DNA molecule unwinds, and each strand serves as a template for the synthesis of a new complementary strand. This process, known as DNA replication, ensures that each daughter cell receives an identical copy of the genetic information.

DNA replication is a crucial process that ensures the accurate duplication of genetic information before cell division. It allows for the transmission of genetic material from parent cells to daughter cells and is essential for the maintenance of genetic stability.

i. **Initiation:** DNA replication begins at specific sites on the DNA molecule called origins of replication. Enzymes called helicases unwind and separate the two DNA strands, forming a replication fork. The separation of the DNA strands creates two template strands for the synthesis of new DNA strands.

ii. **Primer Binding:** Primers, short RNA sequences, are synthesized by the enzyme primase and are complementary to the DNA template strands. Primers serve as the starting points for DNA synthesis. They provide a free 3'-OH group to which DNA polymerase can add new nucleotides.

iii. **Elongation:** DNA polymerase adds new nucleotides to the growing DNA strands in a 5' to 3' direction. However, DNA synthesis can only occur in the 5' to 3' direction, resulting in two different mechanisms at the replication fork:

iv. **Leading Strand:** The leading strand is synthesized continuously in the same direction as the replication fork unwinds. DNA polymerase continuously adds nucleotides to the 3' end of the growing strand, synthesizing it in a smooth and uninterrupted manner.

v. **Lagging Strand:** The lagging strand is synthesized discontinuously, opposite to the direction of the replication fork movement. As the replication fork unwinds, short DNA fragments called Okazaki fragments are synthesized on the lagging strand. DNA polymerase synthesizes each Okazaki fragment separately, starting from the primers. These fragments are later joined by an enzyme called DNA ligase, resulting in a continuous strand.

vi. **Proofreading and Error Correction:** During DNA replication, DNA polymerase has a built-in proofreading mechanism. It checks for errors and incorrect base pairings. If an incorrect base is added, DNA polymerase removes the mismatched nucleotide and replaces it with the correct one before continuing with replication. This proofreading function ensures a high level of accuracy in DNA replication.

vii. **Termination:** DNA replication continues until the replication forks meet at specific termination sites on the DNA molecule. At this point, replication is completed, and two identical DNA molecules have been synthesized. The newly formed DNA molecules are then ready to be segregated into daughter cells during cell division.

Protein Synthesis: Transcription And Translation

Protein Synthesis

Protein synthesis is the process by which cells generate proteins using the genetic information stored in DNA. It involves two main steps: transcription and translation. Let's explore the process of protein synthesis in detail.

1. **Transcription:** Transcription is the first step in protein synthesis and occurs in the nucleus of eukaryotic cells or the cytoplasm of prokaryotic cells. The process can be summarized as follows:

✓ **Initiation:** Transcription begins when an enzyme called RNA polymerase binds to a specific region of DNA called the promoter. The promoter acts as a signal for the start of transcription. Once bound, RNA polymerase unwinds a small section of the DNA double helix, exposing the template strand.

✓ **Elongation:** RNA polymerase adds complementary RNA nucleotides to the growing mRNA (messenger RNA) molecule. It reads the template strand of DNA and synthesizes an RNA molecule that is complementary to the DNA sequence. RNA nucleotides (adenine, uracil, cytosine, and guanine) are added to the growing mRNA chain according to the base-pairing rules (A-U and C-G).

✓ **Termination:** Transcription ends when RNA polymerase reaches a termination signal in the DNA sequence. The termination signal causes RNA polymerase and the newly formed mRNA molecule to detach from the DNA template.

mRNA Processing (in eukaryotes): In eukaryotic cells, the newly synthesized mRNA molecule undergoes additional processing steps before it can be used in translation:

- ✓ **Addition of a 5' Cap:** A modified guanine nucleotide is added to the 5' end of the mRNA molecule. This cap helps protect the mRNA from degradation and assists in the binding of the ribosome during translation.
- ✓ **Polyadenylation:** A poly-A tail, consisting of multiple adenine nucleotides, is added to the 3' end of the mRNA molecule. This tail also aids in mRNA stability and enhances translation efficiency.
- ✓ **Removal of Introns:** Eukaryotic genes often contain non-coding regions called introns. Introns are removed from the pre-mRNA molecule through a process called splicing. The remaining coding regions, called exons, are spliced together to form the mature mRNA molecule.

2. **Translation:** Translation is the second step of protein synthesis and occurs in the cytoplasm, specifically in structures called ribosomes. The process can be divided into three main stages:
- ✓ **Initiation:** The small ribosomal subunit binds to the mRNA molecule at the start codon (usually AUG) with the help of initiation factors. The initiator tRNA, carrying the amino acid methionine, binds to the start codon, positioning the ribosome for protein synthesis.
- ✓ **Elongation:** During elongation, amino acids are added to the growing polypeptide chain. The ribosome moves along the mRNA molecule in a 5' to 3' direction, and each subsequent codon is read by a specific tRNA molecule. The tRNA carries the corresponding amino acid, which is added to the growing polypeptide chain. The ribosome facilitates the formation of peptide bonds between adjacent amino acids.
- ✓ **Termination:** Translation continues until a stop codon (UAA, UAG, or UGA) is reached on the mRNA molecule. When a stop codon enters the ribosome's A site, a release factor

protein binds to the ribosome, causing the polypeptide chain to be released. The ribosome then dissociates from the mRNA molecule.

Protein Folding and Post-Translational Modifications: After translation, the newly synthesized polypeptide chain undergoes folding into its functional three-dimensional structure. Chaperone proteins assist in this process. Additionally, post-translational modifications, such as phosphorylation, glycosylation, or cleavage, may occur to further modify the protein's structure and function.

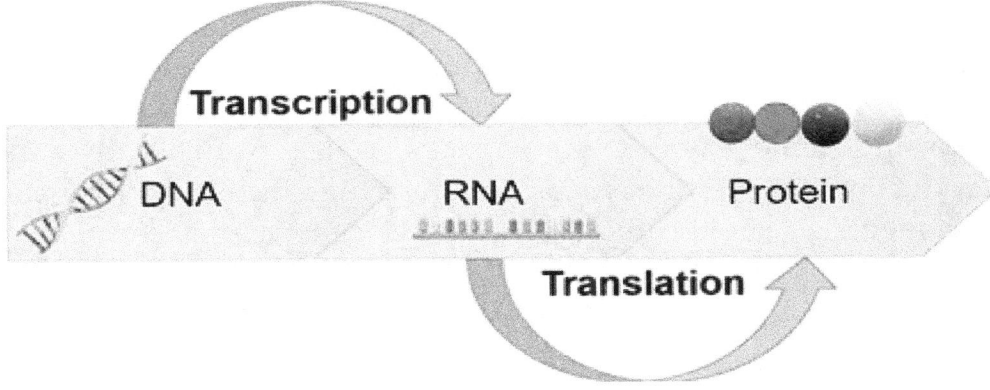

■ Genetic Engineering And Biotechnology

Genetic engineering and biotechnology are two closely related fields that have revolutionized the way we interact with and manipulate living organisms. These disciplines intersect at the molecular level, utilizing genetic information to enhance various processes, improve products, and address various challenges in diverse sectors.

Genetic Engineering

Genetic engineering involves the direct manipulation of an organism's genes using molecular biology techniques. This manipulation can include the introduction, deletion, or alteration of specific genes to achieve desired traits or outcomes. The process often involves the use of recombinant DNA technology, where genetic material from different sources is combined to create a modified organism.

One key application of genetic engineering is the production of genetically modified organisms (GMOs). In agriculture, GMOs are developed to exhibit desirable traits such as resistance to pests, tolerance to herbicides, or enhanced nutritional content. This can lead to increased crop yields and reduced reliance on chemical inputs. In medicine, genetic engineering has paved the way for gene therapy, a promising field aimed at treating genetic disorders by introducing or repairing genes within a patient's cells. Additionally, the production of recombinant proteins, such as insulin for diabetes treatment, is a significant achievement of genetic engineering.

Biotechnology

Biotechnology is a broader field that encompasses the application of biological systems, organisms, or derivatives to develop or create new products or processes. Genetic engineering is just one aspect of biotechnology, and the field includes various other techniques and applications.

Applications Of Biotechnology

1. **Medicine:** Biotechnology has revolutionized the pharmaceutical industry. It includes the production of vaccines, antibiotics, and therapeutic proteins through the use of genetically engineered microorganisms. Personalized

medicine, based on an individual's genetic makeup, is also an emerging frontier.
2. **Agriculture:** Besides genetic engineering, biotechnology in agriculture includes techniques like selective breeding, tissue culture, and marker-assisted selection to improve crop yields, develop disease-resistant varieties, and create plants with enhanced nutritional profiles.
3. **Environmental Management:** Biotechnology plays a crucial role in environmental protection and resource management. It includes the use of microorganisms to degrade pollutants, clean up oil spills, and enhance soil fertility.
4. **Industrial Processes:** Biotechnological processes are employed in various industries, including food and beverage production, textile manufacturing, and biofuel production. Enzymes and microorganisms are utilized to enhance production processes and reduce environmental impact.
5. **Forensics:** Biotechnology is employed in forensic science for DNA profiling, aiding in criminal investigations and paternity testing.
6. **Bioinformatics:** Advances in computing technology and biology have led to the development of bioinformatics, a field that involves the use of computational tools to analyze biological data, study genetic sequences, and understand complex biological systems.

EVOLUTION AND NATURAL SELECTION — CHAPTER 4

■ The Theory Of Evolution

The Theory of Evolution is a scientific explanation for the diversity of life on Earth and the processes that have shaped it over time. Proposed by Charles Darwin in the mid-19th century, it revolutionized our understanding of the natural world and remains one of the most fundamental concepts in biology. The theory of evolution is supported by a wide range of evidence from various scientific disciplines, including paleontology, genetics, comparative anatomy, and molecular biology.

The theory of evolution proposes that all living organisms share a common ancestor and have descended from it through a process of modification. Over millions of years, populations of organisms have undergone changes, resulting in the development of new species. This concept emphasizes the unity of life and the idea that all organisms are part of a vast tree of life.

Natural Selection

Natural selection is the primary mechanism driving evolutionary change. It is based on the following principles:

a. **Variation:** Individuals within a population exhibit variation in their traits, which is influenced by genetic differences and environmental factors.

b. **Overproduction:** Populations have the potential to produce more offspring than the environment can support, leading to competition for resources.
c. **Differential Survival and Reproduction:** Individuals with traits that are advantageous for survival and reproduction have a higher likelihood of surviving and passing on their genes to the next generation.
d. **Heredity:** Offspring inherit traits from their parents, passing on the advantageous traits to future generations.
e. **Gradualism and Speciation:** The theory of evolution suggests that evolutionary changes occur gradually over long periods. Small modifications in traits accumulate over generations, leading to significant changes in organisms over time. This gradual process can ultimately result in the formation of new species through a process known as speciation. Speciation occurs when populations become reproductively isolated from each other, preventing gene flow and allowing them to diverge genetically.

■ Evidence For Evolution

The theory of evolution is supported by a vast array of evidence from various scientific disciplines. This evidence comes from the fields of paleontology, comparative anatomy, embryology, molecular biology, and biogeography. The cumulative evidence from these and many other sources strongly supports the theory of evolution as the best explanation for the diversity and interconnectedness of life on Earth. It provides a unifying framework that is widely accepted by the scientific community. Here are some of the key evidence for evolution:

1. **Fossil Record:** Fossils provide a record of past life forms and their evolutionary changes over time. The fossil record shows a progression of increasingly complex organisms and the existence of transitional forms that bridge gaps between

different species. Examples include the fossils of ancient fish with primitive limbs, indicating the transition from aquatic to terrestrial life.
2. **Comparative Anatomy:** Comparative anatomy studies the similarities and differences in the structures of different organisms. Homologous structures, which have the same underlying structure but may have different functions, suggest a common ancestry. For example, the pentadactyl limb (five-digit limb) is found in a wide range of vertebrates, including humans, whales, bats, and birds, suggesting a shared evolutionary history.
3. **Embryology:** Comparative embryology examines the development of embryos across different species. Many organisms show similar embryonic stages and structures, indicating shared ancestry. For example, the early embryos of fish, reptiles, birds, and mammals have gill slits and tails, reflecting their common aquatic ancestry.
4. **Molecular Biology:** Molecular biology provides strong evidence for evolution. DNA and protein sequences can be compared across different species to determine their degree of relatedness. The more similar the sequences, the more closely related the species are believed to be. The universality of the genetic code across organisms further supports the idea of a common ancestry.
5. **Biogeography:** Biogeography studies the distribution of species across different geographic regions. It provides evidence for how species have evolved and dispersed over time. For example, the similarities between species on nearby islands suggest that they share a common ancestor and have diverged due to geographic isolation.
6. **Experimental Evolution:** Experimental evolution involves studying evolutionary processes in real-time using laboratory experiments. Researchers can observe changes in organisms

over generations, such as the development of antibiotic resistance in bacteria or the evolution of new traits in fruit flies. These experiments provide direct evidence for the mechanisms and rates of evolutionary change.
7. **Transitional Fossils:** Transitional fossils are fossilized remains of organisms that exhibit characteristics of both ancestral and descendant groups. These fossils provide direct evidence for the gradual transitions and intermediate stages in the evolution of different species. Well-known examples include Tiktaalik, a fish-like creature with limb-like fins, considered an intermediate form between fish and tetrapods.
8. **Convergent Evolution:** Convergent evolution occurs when unrelated species independently evolve similar traits or adaptations due to similar environmental pressures. Examples include the wings of bats and birds or the streamlined bodies of dolphins and sharks. Convergent evolution highlights the role of natural selection in shaping similar traits in different lineages.

■ Mechanisms Of Evolution

The mechanisms of evolution are the processes that drive changes in the genetic composition of populations over time. These mechanisms can lead to the emergence of new species, the adaptation of organisms to their environments, and the overall diversity of life. The main mechanisms of evolution are as follows:
1. **Natural Selection:** Natural selection is the most well-known and powerful mechanism of evolution. It acts on the variation that exists within populations. The key components of natural selection are:
i. **Variation:** Individuals within a population exhibit variation in their traits, which can be influenced by genetic differences, mutations, and environmental factors.

ii. **Selective Pressure:** The environment exerts pressure on organisms, favoring certain traits that enhance survival and reproductive success.
iii. **Differential Reproduction:** Individuals with traits that are advantageous in the given environment have a higher chance of surviving and reproducing, passing on their favorable traits to the next generation.
iv. **Heredity:** Offspring inherit traits from their parents, including the advantageous traits that were selected for. This leads to an increase in the frequency of beneficial traits in the population over time.
v. Natural selection can result in adaptations, where the traits that confer a selective advantage become more prevalent in the population, enhancing the organism's fitness for its environment.

2. **Genetic Drift:** Genetic drift refers to random fluctuations in the frequency of alleles (alternative forms of genes) within a population. It is driven by chance events rather than selective pressures. Genetic drift can have a more significant impact on small populations where chance events can lead to substantial changes in allele frequencies over generations. Genetic drift can lead to the loss of certain alleles from a population (known as genetic bottleneck) or the fixation of an allele where it becomes the only variant present (known as genetic fixation).=

3. **Gene Flow:** Gene flow occurs when individuals or their genes move between populations. It can happen through migration, interbreeding, or the movement of seeds or spores. Gene flow can introduce new genetic variation into a population and reduce genetic differences between populations. It plays a crucial role in maintaining genetic diversity within a species and counteracting the effects of genetic drift and natural selection.

4. **Mutation:** Mutations are spontaneous changes in the DNA sequence of an organism's genome. They are the ultimate source of genetic variation in populations. Mutations can be caused by various factors, such as errors in DNA replication, exposure to radiation, or chemical mutagens. Most mutations are neutral or harmful, but occasionally they can be beneficial and provide an advantage in certain environments. Beneficial mutations can be acted upon by natural selection, leading to their increased frequency in a population over time.
5. **Genetic Recombination:** Genetic recombination occurs during sexual reproduction when genetic material from two parents combines to create new combinations of alleles. It happens through processes like crossing over and independent assortment during meiosis. Genetic recombination shuffles existing genetic variation, potentially creating novel combinations of alleles that can be subject to natural selection.

■ Natural Selection

Natural selection is a fundamental mechanism of evolution proposed by Charles Darwin. It is the process by which certain heritable traits become more or less common in a population over successive generations based on their impact on an organism's survival and reproductive success. Natural selection is not a conscious or deliberate process. It is the result of the interactions between individuals, their traits, and the environment. Over time, natural selection shapes populations, allowing them to become better suited to their respective environments and driving the ongoing diversification of life on Earth. It acts on the variation that exists within populations and is driven by the following

i. **Variation:** Individuals within a population exhibit variation in their traits, which can be influenced by genetic differences, mutations, and environmental factors. This variation can

include differences in physical characteristics, behaviors, physiological processes, or even genetic makeup.
ii. **Selective Pressure:** The environment exerts pressure on organisms, favoring certain traits over others. Selective pressures can include factors such as predation, competition for resources, climate conditions, availability of food, and mate selection. These pressures create challenges and opportunities for individuals with different traits to survive and reproduce.
iii. **Differential Reproduction:** Individuals with traits that are advantageous in the given environment have a higher chance of surviving and reproducing successfully, thus passing on their favorable traits to the next generation. This leads to differential reproduction, where individuals with advantageous traits leave more offspring compared to those with less advantageous or detrimental traits.
iv. **Heredity:** Offspring inherit traits from their parents through the transmission of genetic information. This means that the traits that were selected for in one generation are more likely to be present in the next generation. Over time, as advantageous traits are passed on and accumulated through successive generations, they become more prevalent in the population.

The Process Of Natural Selection Can Lead To Several Outcomes

1. **Adaptation:** Natural selection acts on variation and favors traits that enhance an organism's fitness for its environment. This process of adaptation allows individuals with advantageous traits to be better equipped to survive, reproduce, and pass on those traits to future generations.

Adaptations can be structural, physiological, or behavioral, and they increase an organism's chances of surviving and reproducing in its specific environment.
2. **Evolutionary Change:** Over long periods, natural selection can result in significant evolutionary changes within a population or lead to the formation of new species. Through the accumulation of advantageous traits and the elimination of less favorable ones, populations can diverge and give rise to new lineages.
3. **Extinction:** Natural selection can also lead to the extinction of less adapted or less fit species. If a species is unable to cope with changing environmental conditions or is outcompeted by better-adapted organisms, it may become extinct.

■ Speciation And Adaptive Radiation

Speciation and adaptive radiation are key processes in the evolutionary history of organisms. They involve the formation of new species and the subsequent diversification of those species into different ecological niches. Let's explore these concepts in detail:

Speciation

Speciation is the process by which new species arise from existing populations. It occurs when populations become reproductively isolated from each other, meaning they can no longer interbreed and produce fertile offspring. There are two primary modes of speciation:
1. **Allopatric Speciation:** Allopatric speciation occurs when populations become geographically isolated from each other. Geographic barriers, such as mountains, bodies of water, or other forms of physical separation, prevent gene flow between the populations. Over time, genetic differences can accumulate due to mutation, genetic drift, and natural

selection. Eventually, the populations become distinct species with unique genetic and phenotypic characteristics.
2. **Sympatric Speciation:** Sympatric speciation occurs when populations diverge into separate species within the same geographic area, without physical barriers. This process often involves mechanisms that reduce gene flow between different subgroups within a population. These mechanisms can include polyploidy (changes in chromosome number), disruptive selection (favoring different phenotypes), or assortative mating (preference for mating with individuals with similar traits). Sympatric speciation is more common in plants than in animals.

Adaptive Radiation

Adaptive radiation refers to the rapid diversification of a common ancestor into a wide array of species, each adapted to different ecological niches. It typically occurs when a lineage colonizes a new and diverse environment with abundant available resources and relatively few existing competitors. Adaptive radiation can lead to the emergence of various species with distinct traits, adaptations, and ecological roles.

The process of adaptive radiation involves the following steps:

i. **Colonization:** A small group of organisms from a single species colonizes a new and unoccupied environment or ecological niche. This colonization event can occur due to dispersal or a change in habitat.

ii. **Divergent Selection:** Within the new environment, different ecological opportunities and selective pressures arise. This leads to divergent selection, favoring different traits or adaptations in different subpopulations. Over time, this divergence results in the formation of new species.

iii. **Ecological Niches:** The newly formed species occupy different ecological niches within the environment, exploiting specific

resources or habitats. Each species evolves unique adaptations and characteristics that allow them to thrive in their specific niche.
iv. **Coexistence:** The newly evolved species coexist in the same geographical area, occupying different ecological niches and minimizing competition for resources. This coexistence promotes further divergence and specialization.

Examples of adaptive radiation include the finches of the Galapagos Islands, where different beak shapes evolved in response to varying food sources, and the cichlid fish in the African Great Lakes, which display remarkable diversity in morphology and feeding strategies.

■ Human Evolution

Human evolution is the evolutionary process that led to the emergence of modern humans, Homo sapiens, from ancestral primates. The study of human evolution combines evidence from paleontology, comparative anatomy, genetics, and archaeological discoveries.

- ➤ **Hominin Lineage:** The hominin lineage includes modern humans and our closest extinct relatives. The earliest hominin species appeared in Africa around 6-7 million years ago. Over time, various species evolved, displaying a progression of anatomical and behavioral changes.
- ➤ **Australopithecus:** The genus Australopithecus, which includes famous species like Australopithecus afarensis (represented by the fossil "Lucy"), lived around 4-2 million years ago. Australopithecus species had a mix of ape-like and human-like traits. They were bipedal (walked on two legs) and showed some evidence of tool use.
- ➤ **Genus Homo:** The genus Homo emerged around 2.8 million years ago with Homo habilis, an early tool-maker. Homo

erectus, a species that lived around 1.8 million to 200,000 years ago, was the first to widely disperse beyond Africa and had a more sophisticated tool-making culture.
- **Neanderthals:** Neanderthals, or Homo neanderthalensis, were a distinct human species that inhabited Europe and parts of Asia from around 400,000 to 40,000 years ago. They shared a common ancestor with modern humans and had a robust physique and larger brains. Genetic evidence suggests that there was some interbreeding between Neanderthals and modern humans.
- **Modern Humans:** Modern humans, Homo sapiens, emerged in Africa around 300,000 years ago. They gradually migrated across the globe, eventually replacing other hominin species, including the Neanderthals. Modern humans have distinctive features such as a high forehead, smaller face and teeth, and a more complex cognitive capacity.

The human lineage

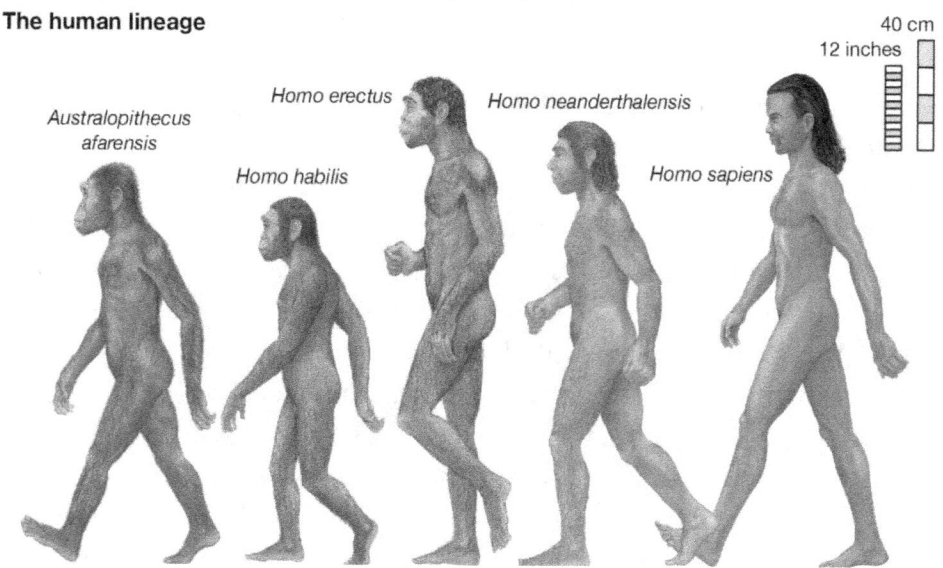

Cultural Evolution: Alongside biological changes, human evolution also involves cultural evolution, where knowledge, tools, language, and social systems developed and were transmitted across generations. Cultural evolution played a crucial role in human adaptation and success.

Genetic Studies: Genetic studies have provided valuable insights into human evolution. Comparisons of DNA from different human populations around the world reveal patterns of migration, interbreeding, and common ancestry. These studies have helped reconstruct the human family tree and understand our shared genetic heritage.

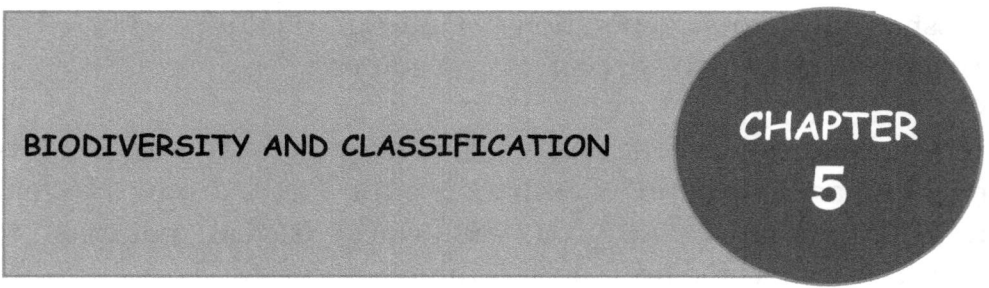

Introduction To Biodiversity

Biodiversity refers to the variety of life on Earth at all levels of organization, including genetic, species, and ecosystem diversity. It encompasses the millions of species of plants, animals, microorganisms, and the ecosystems they form. Biodiversity is a fundamental aspect of our planet's natural heritage and is essential for the functioning of ecosystems and the well-being of all living organisms, including humans. Here are some key points to understand the concept of biodiversity:

Levels Of Biodiversity

i. **Genetic Diversity:** Genetic diversity refers to the variety of genes within a species. It is essential for the adaptation and survival of populations, as it provides the raw material for evolutionary processes. Genetic diversity allows species to respond to changes in their environment, such as climate change or new diseases.

ii. **Species Diversity:** Species diversity refers to the variety of different species present in a particular region or on Earth as a whole. It encompasses the richness (number of species) and evenness (relative abundance of species) of species in a given area. High species diversity is an indicator of a healthy and resilient ecosystem.

iii. **Ecosystem Diversity:** Ecosystem diversity refers to the variety of ecosystems present in a region or on Earth. Ecosystems are dynamic communities of organisms interacting with their environment. They can range from forests and grasslands to coral reefs and wetlands. Each ecosystem has unique physical and biological characteristics, and their diversity ensures the stability and functioning of the Earth's biosphere.

Importance Of Biodiversity

1. **Ecosystem Services:** Biodiversity plays a crucial role in providing numerous ecosystem services that are essential for human well-being. These services include the purification of air and water, nutrient cycling, pollination of crops, regulation of climate, and provision of food, medicine, and raw materials.
2. **Resilience and Adaptation:** Biodiversity enhances the resilience of ecosystems and species in the face of environmental changes and disturbances. Diverse ecosystems are better able to withstand and recover from disturbances such as natural disasters, disease outbreaks, or climate change.
3. **Economic Value:** Biodiversity contributes to the economy through sectors such as agriculture, forestry, fisheries, and tourism. It provides employment opportunities, food security, and livelihoods for many communities around the world.
4. **Aesthetic and Cultural Value:** Biodiversity also has intrinsic value, providing aesthetic pleasure, inspiration, and cultural significance. It forms the basis of cultural practices, traditional knowledge, and spiritual beliefs of many indigenous communities.

Threats To Biodiversity

1. **Habitat Loss and Fragmentation:** The destruction and fragmentation of natural habitats due to activities such as deforestation, urbanization, and conversion of land for agriculture are major threats to biodiversity. Loss of habitat disrupts ecosystems, leads to the extinction of species, and reduces overall biodiversity.
2. **Climate Change:** Climate change poses significant challenges to biodiversity. Rising temperatures, altered precipitation patterns, and extreme weather events can disrupt ecosystems, cause shifts in species' distribution, and result in the loss of vulnerable species and habitats.
3. **Pollution and Contamination:** Pollution, including air and water pollution, as well as the release of chemicals and pollutants into the environment, can have detrimental effects on biodiversity. It can lead to the decline of species, disruption of ecosystems, and ecological imbalances.
4. **Overexploitation:** Unsustainable harvesting of species for food, medicine, trade, and other purposes can push species to extinction and disrupt ecosystems. Overfishing, illegal wildlife trade, and poaching are examples of activities that threaten biodiversity.

Conservation And Sustainable Use

Protecting and conserving biodiversity is crucial for the long-term sustainability of our planet. Conservation efforts involve establishing protected areas, implementing sustainable resource management practices, promoting sustainable agriculture, reducing pollution, and raising awareness about the value of biodiversity.

Additionally, sustainable use of biodiversity ensures that we can meet our current needs while preserving resources for future generations. This involves adopting practices that minimize the impact on ecosystems, promoting responsible consumption, and supporting sustainable livelihoods for local communities.

■ Classification and Taxonomy

Classification and taxonomy are branches of biology that involve the organization and categorization of living organisms based on their similarities and evolutionary relationships. These fields provide a systematic way to understand the diversity of life on Earth and establish a common language for scientists to communicate and study biological organisms. Here are some terms relating to Classification and Taxonomy

➤ **Classification:** Classification is the process of arranging organisms into groups or categories based on shared characteristics. It involves grouping organisms into various levels of hierarchy to reflect their evolutionary relationships. The main purpose of classification is to organize and categorize the vast number of species on Earth, making it easier to study and understand the relationships between different organisms.

➤ **Taxonomy:** Taxonomy is the scientific discipline within biology that focuses on identifying, describing, and classifying organisms. It encompasses the principles, methods, and rules for naming and organizing organisms into a hierarchical system. Taxonomists use a combination of morphological, anatomical, genetic, and ecological characteristics to classify organisms and determine their relationships.

➤ **Hierarchical Structure:** Taxonomy follows a hierarchical structure that consists of several levels, or taxa, arranged in a nested pattern. The hierarchy, from broadest to most specific, includes domains, kingdoms, phyla (or divisions for plants), classes, orders, families, genera, and species. Each level represents a different degree of relatedness among organisms, with the species level being the most specific.

- **Binomial Nomenclature:** Binomial nomenclature is the naming system used in taxonomy to assign a unique scientific name to each species. It involves using two Latinized names: the genus (capitalized) and the species epithet (lowercase). The scientific name is written in italics or underlined. For example, Homo sapiens is the scientific name for modern humans, with Homo representing the genus and sapiens representing the species epithet.
- **Phylogenetics:** Phylogenetics is a field within taxonomy that focuses on reconstructing the evolutionary relationships among organisms. It uses various data sources, including molecular data, morphology, and fossil records, to construct phylogenetic trees or cladograms. These trees depict the branching patterns and relatedness of organisms, helping to determine their evolutionary history.
- **Modern Taxonomy:** Modern taxonomy incorporates not only morphological characteristics but also genetic and molecular data to establish evolutionary relationships accurately. Advances in DNA sequencing techniques have greatly improved the understanding of relationships among organisms by providing insights into genetic similarities and differences.
- **Taxonomic Ranks:** In addition to the main hierarchical levels, taxonomists use additional ranks to further classify organisms within each level. These ranks include subphylum, superclass, subclass, superorder, subfamily, subgenus, and others. The specific ranks used may vary depending on the group of organisms being classified.

Introduction To Biology

Kingdoms Of Life

The kingdoms of life are broad categories used in biological classification to classify and organize living organisms based on their characteristics, evolutionary relationships, and complexity. Classification of organisms is constantly evolving as new scientific discoveries are made, the development of molecular techniques and advances in DNA sequencing have led to the proposal of alternative classification systems and the identification of new groups. However, the six kingdoms provide a general framework for understanding the diversity of life on Earth and serve as a basis for further classification and study of organisms. Here is an explanation of the six kingdoms:

1. **Kingdom Animalia:** The kingdom Animalia consists of multicellular organisms known as animals. Animals are characterized by their ability to move, obtain energy by consuming other organisms, and lack cell walls. They exhibit a wide range of characteristics, body plans, and behaviors, and encompass a diverse array of species from microscopic invertebrates to large mammals.
2. **Kingdom Plantae:** The kingdom Plantae includes multicellular organisms known as plants. Plants are characterized by their ability to perform photosynthesis, have cell walls composed of cellulose, and typically remain stationary. They play a crucial role in ecosystems as primary producers, converting sunlight, water, and carbon dioxide into energy-rich molecules.
3. **Kingdom Fungi:** The kingdom Fungi comprises organisms known as fungi. Fungi are characterized by their ability to obtain nutrients by decomposing organic matter or living as parasites on other organisms. They have cell walls made of chitin and include familiar organisms such as mushrooms, yeasts, and molds.
4. **Kingdom Protista:** The kingdom Protista consists of diverse eukaryotic organisms that do not fit into the other kingdoms.

Protists are mostly unicellular, although some can be multicellular. They display a wide variety of forms, lifestyles, and reproductive strategies. Protists include organisms such as amoebas, algae, and protozoa.
5. **Kingdom Archaea:** The kingdom Archaea comprises a group of single-celled microorganisms. Archaea are prokaryotes and are distinct from bacteria (which belong to the kingdom Bacteria) and eukaryotes. They are known for their ability to thrive in extreme environments such as hot springs, acidic environments, and deep-sea hydrothermal vents.
6. **Kingdom Bacteria:** The kingdom Bacteria consists of prokaryotic microorganisms known as bacteria. Bacteria are found in a wide range of habitats and display diverse metabolic capabilities. Some bacteria are beneficial, such as those involved in nutrient cycling and nitrogen fixation, while others can be pathogenic and cause diseases.

Note: Initially both Kingdom Archaea and kingdom bacteria were classified as Monerans

Animal Kingdom

R.H. Whittaker organized organisms into five kingdoms. He classified organisms based on cell structure, mode and source of nutrition and body design. The five kingdoms proposed by Whittaker are Monera, Protista, Fungi, Plantae and Animalia. Let us learn about the animal kingdom, i.e., Kingdom Animalia.

Kingdom Animalia

Kingdom Animalia constitutes all animals. Amongst the five kingdoms, the largest kingdom is the animal kingdom.

Animals are multicellular eukaryotes. However, unlike plants, they do not possess chlorophyll or a cell wall. Therefore, members of the animal kingdom exhibit a heterotrophic mode of nutrition. Kingdom Animalia has been classified into ten different subphyla based on their body design or differentiation.
The different phylum of the animal kingdom are as follows:
Porifera, Coelenterata (Cnidaria), Platyhelminthes, Nematoda, Annelida, Arthropoda, Mollusca, Echinodermata, Hemichordata, Chordata

Phylum Porifera
Porifera means organisms with holes. They are commonly known as Sponges. Features of the poriferan are:
i. **Cellular Organization:** Porifera have a cellular level of organization, meaning their bodies are made up of specialized cells but lack true tissues and organs. They have a loose aggregation of cells that work together to form a functional body.
ii. **Porous Body Structure:** Sponges have a porous body structure characterized by numerous small openings called ostia. These ostia allow water to enter the sponge and flow through a system of internal canals.
iii. **Choanocytes:** Choanocytes, also known as collar cells, are a distinctive type of cells found in sponges. They have a collar-like structure with a flagellum surrounded by a collar of microvilli. Choanocytes generate water currents and help in capturing food particles.
iv. **Spicules and Spongin:** Sponges often possess structural elements called spicules, which are tiny, needle-like or spiny structures made of calcium carbonate or silica. Some sponges also have a flexible protein-based framework called spongin, which provides support to the sponge body.

v. **Filter Feeding:** Sponges are filter feeders, obtaining their nutrition by filtering small particles, such as bacteria and organic matter, from the water that passes through their bodies. The choanocytes create water currents that facilitate the capture of food particles.
vi. **Asexual Reproduction:** Sponges can reproduce asexually through a process called budding. Budding involves the formation of a small outgrowth, or bud, from the parent sponge, which eventually grows and develops into a new individual.
vii. **Sexual Reproduction:** Sponges also reproduce sexually. They produce eggs and sperm, which are released into the water, allowing for external fertilization. The fertilized eggs develop into free-swimming larvae that disperse and settle in new locations to establish new sponge colonies.
viii. **Regeneration Ability:** Sponges have a remarkable ability to regenerate. If a sponge is fragmented or damaged, the cells have the capacity to reorganize and rebuild the damaged parts, ultimately restoring the structure and functionality of the sponge.
ix. **Ecological Role:** Sponges play important ecological roles in marine ecosystems. They provide habitats for other organisms, serving as shelter and nursery areas. Sponges also contribute to nutrient cycling by filtering water and recycling organic matter.

Examples of phylum Porifera include- Spongilla, Sycon.

©2001 California Academy of Sciences

Phylum Coelenterata (Cnidaria)

i. **Radial Symmetry:** Cnidarians typically exhibit radial symmetry, which means their body parts are arranged around a central axis, allowing them to be divided into similar halves in multiple planes.

ii. **Tissue-Level Organization:** Cnidarians have a tissue-level organization, meaning their bodies are composed of different types of specialized tissues but lack complex organs and organ systems.

iii. **Two Body Forms:** Cnidarians exist in two primary body forms:
✓ **Polyp:** The polyp form is cylindrical and typically sessile, attached to a substrate. It has a cylindrical body with a central mouth surrounded by tentacles.

- ✓ **Medusa:** The medusa form is bell-shaped and free-swimming. It has a gelatinous body with a central mouth and tentacles hanging down from the edge.
iv. **Tentacles and Cnidocytes:** Cnidarians have tentacles, which are armed with specialized stinging cells called cnidocytes. Cnidocytes contain nematocysts, tiny harpoon-like structures that can inject venom into prey or potential threats.
v. **Gastrovascular Cavity:** Cnidarians possess a central digestive cavity called the gastrovascular cavity. This cavity serves for both digestion and distribution of nutrients throughout the body.
vi. **Nerve Net:** Cnidarians have a simple nervous system composed of a decentralized network of neurons called a nerve net. This allows for coordination of basic sensory and motor functions.
vii. **Reproduction:** Cnidarians can reproduce both sexually and asexually. They can produce eggs and sperm that are released into the water for external fertilization. Some species also exhibit asexual reproduction through budding or regeneration.
viii. **Examples of Cnidarians:** Phylum Cnidaria includes various familiar organisms, such as jellyfish, sea anemones, corals, and hydroids. Examples of cnidarian species include the moon jellyfish (Aurelia aurita), the lion's mane jellyfish (Cyanea capillata), the sea anemone (Actinia spp.), and the reef-building coral (Acropora spp).

Introduction To Biology

Aurelia aurita; Moon Jellyfish
©2002 California Academy of Sciences

Phylum Platyhelminthes

Platyhelminthes are commonly known as flatworms. Their features are:

i. **Bilateral Symmetry:** Flatworms exhibit bilateral symmetry, meaning their bodies can be divided into two similar halves along a single plane.

ii. **Flattened Body Shape:** As their name suggests, flatworms have a thin and flattened body shape, which allows them to occupy tight spaces and move easily through their environments.

iii. **Three Primary Germ Layers:** Platyhelminthes have three primary germ layers during embryonic development, known as triploblastic organization. These layers give rise to different tissues and organ systems in the body.

iv. **Cephalization:** Many flatworms show a concentration of sensory organs and nerve cells at their anterior end, a

characteristic known as cephalization. This allows for better detection of stimuli and more complex sensory processing.
v. **Acoelomate Body Plan:** Flatworms are acoelomates, which means they lack a true body cavity called a coelom. Instead, their organs are embedded in a solid, mesenchyme-filled tissue.
vi. **Digestive System:** Platyhelminthes exhibit a single opening for both ingestion of food and elimination of waste. This means they have a gastrovascular cavity, which functions for digestion and distribution of nutrients.
vii. **Reproduction:** Flatworms can reproduce both sexually and asexually. Some species are hermaphroditic, possessing both male and female reproductive organs, while others have separate sexes. They may undergo internal or external fertilization, depending on the species.
viii. **Regeneration:** Many flatworms have the remarkable ability to regenerate lost body parts, including the head or tail region. This regenerative capacity is due to the presence of pluripotent cells in their bodies.
ix. **Parasitic Lifestyles:** Some flatworms are parasitic, living in or on other organisms and obtaining nutrients from their hosts. Examples include tapeworms and flukes, which have complex life cycles involving multiple hosts.
x. **Habitat Diversity:** Platyhelminthes can be found in various habitats, including freshwater, marine environments, and moist terrestrial habitats. They can be free-living, symbiotic, or parasitic, depending on the species.

Examples of organisms in Phylum Platyhelminthes:
- Planarians (Class Turbellaria),
- Tapeworms (Class Cestoda),
- Flukes (Class Trematoda),
- Monogeneans (Class Monogenea)

Introduction To Biology

Taenia solium

Phylum Nematoda

The phylum Nematoda, commonly known as roundworms, encompasses a diverse group of organisms with distinct features. Here are some general characteristics and features of Nematoda:

i. **Body Structure:** Nematodes have a cylindrical, elongated body with a tapered, rounded shape at both ends. They are unsegmented, meaning they do not have distinct body segments.

ii. **Bilateral Symmetry:** Like most animals, nematodes exhibit bilateral symmetry, meaning their bodies can be divided into two equal halves along a central axis.

iii. **Pseudocoelomate Body Cavity:** Nematodes are pseudocoelomates, which means they have a body cavity called a pseudocoelom. The pseudocoelom is a fluid-filled cavity located between the mesoderm and endoderm.

iv. **Cuticle:** Nematodes have a tough, flexible outer covering called a cuticle. The cuticle is secreted by the epidermis and provides support and protection for the body. It also plays a role in locomotion.

v. **Digestive System:** Nematodes have a complete digestive system, with a mouth at the anterior end and an anus at the posterior end. The digestive system includes a muscular pharynx for feeding and the intestine for digestion and absorption of nutrients.

vi. **Reproductive System:** Nematodes have separate sexes, with males and females having distinct reproductive organs. Some nematodes are also capable of asexual reproduction through parthenogenesis.

vii. **Muscular System:** Nematodes have longitudinal muscles that run along the length of their body. These muscles contract and relax, enabling nematodes to move and burrow through various substrates.

viii. **Sensory Structures:** Nematodes have sensory structures such as nerve rings and sensory bristles that help them detect environmental cues. They also possess specialized sense organs, including chemoreceptors, mechanoreceptors, and photoreceptors.

ix. **Diversity and Adaptations:** Nematodes are incredibly diverse and can be found in almost every habitat on Earth. They have adapted to a wide range of ecological niches, including soil, freshwater, marine environments, and even as parasites of plants, animals, and humans.

Examples of organisms in Phylum Nematoda: Caenorhabditis elegans, Ascaris lumbricoides, Trichinella spiralis, Hookworms (Ancylostoma spp. and Necator americanus)

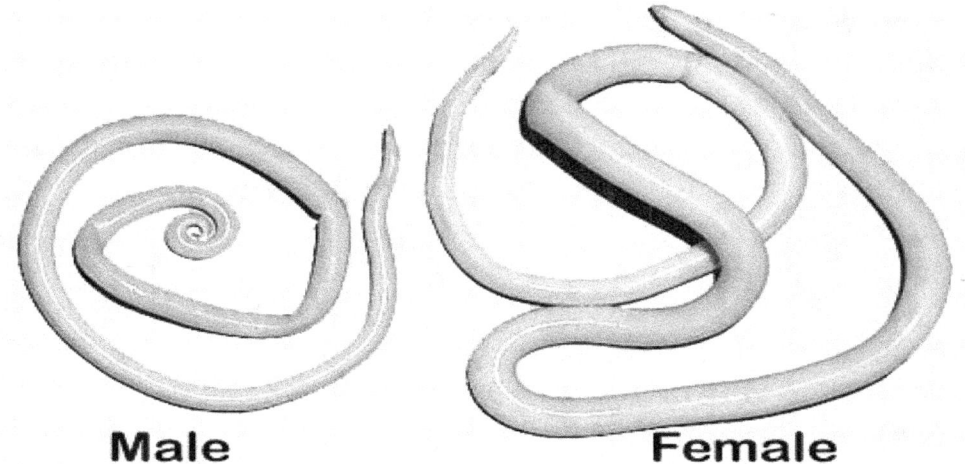

Ascaris lumbricoides

Phylum Annelida

Phylum Annelida comprises segmented worms, commonly known as annelids.

i. **Segmentation:** Annelids exhibit a highly characteristic feature called metamerism, which is the presence of distinct body segments along the length of the animal. Each segment contains a set of repeated organs and structures.

ii. **Coelom:** Annelids have a true coelom, which is a fluid-filled body cavity lined with mesoderm. The coelom provides space for internal organs and allows for greater complexity and specialization within the body.

iii. **Body Wall:** The body wall of annelids consists of three layers: an outer circular muscle layer, an inner longitudinal muscle layer, and a thin cuticle or epidermis. The coordinated contraction and relaxation of these muscle layers enable movement and locomotion.

iv. **Chaetae:** Most annelids possess bristle-like structures called chaetae or setae, which project from the body wall. Chaetae provide traction and assist in locomotion, as well as serve as sensory organs.

v. **Complete Digestive System:** Annelids have a complete digestive system with a mouth and an anus. The digestive tract consists of specialized regions, such as the pharynx, esophagus, crop, gizzard, and intestine, which are involved in ingestion, digestion, and absorption of nutrients.
vi. **Closed Circulatory System:** Annelids have a closed circulatory system composed of blood vessels that transport oxygen, nutrients, and waste products throughout the body. The circulatory system typically consists of a dorsal blood vessel, ventral blood vessels, and smaller lateral vessels.
vii. **Nervous System:** Annelids have a relatively well-developed nervous system. They possess a ventral nerve cord that runs along the length of the body and is connected to a collection of ganglia (clusters of nerve cell bodies) in each body segment. The nerve cord helps coordinate movement and sensory responses.
viii. **Respiration:** Annelids employ various respiratory mechanisms depending on the species. Some species exchange gases through their body surface, while others have specialized respiratory structures, such as gills or thin-walled regions of the body.

Examples of annelids include: Earthworm(Class Oligochaeta), Polychaetes (Class Polychaeta), Leeches (Class Hirudinea)

Eathworm

Phylum Arthropoda

Arthropoda is a vast phylum that includes a diverse group of invertebrate animals. Here are some general features of arthropods:

i. **Segmented Body:** Arthropods have a segmented body organized into distinct regions, including a head, thorax, and abdomen (in most species). The body is often covered by a hard exoskeleton made of chitin, which provides protection and support.

ii. **Jointed Appendages:** Arthropods possess jointed appendages, such as legs, antennae, and mouthparts. These appendages allow for a wide range of movement, manipulation of objects, sensory perception, and feeding.

iii. **Bilateral Symmetry:** Arthropods exhibit bilateral symmetry, meaning their body can be divided into two equal halves along a central plane.

iv. **Exoskeleton:** Arthropods have an external exoskeleton composed of chitin, a tough and flexible protein. The exoskeleton provides support, protection against predators, and serves as a site for muscle attachment. However, arthropods must molt or shed their exoskeleton periodically to grow.

v. **Well-Developed Nervous System:** Arthropods have a relatively advanced nervous system. They possess a brain located in the head region and a ventral nerve cord that extends throughout the body. Arthropods also have various sensory structures, including compound eyes and antennae, which allow for a range of sensory perceptions.

vi. **Open Circulatory System:** Most arthropods have an open circulatory system, where a fluid called hemolymph is pumped through the body cavity, called a hemocoel, by a muscular heart. Hemolymph plays a role in nutrient distribution, waste removal, and immune response.

vii. **Metamorphosis:** Many arthropods undergo metamorphosis during their life cycle, which involves distinct developmental stages. Examples of metamorphosis include complete metamorphosis (holometabolous) seen in insects like butterflies, where individuals pass through egg, larval, pupal, and adult stages, and incomplete metamorphosis (hemimetabolous) seen in insects like grasshoppers, where individuals progress through egg, nymph, and adult stages.

viii. **High Species Diversity:** Arthropods are the most diverse animal phylum, with over one million described species. They inhabit various terrestrial, freshwater, and marine ecosystems, occupying a wide range of niches and ecological roles.

Examples of arthropods include insects (e.g., beetles, butterflies, ants), arachnids (e.g., spiders, scorpions), crustaceans (e.g., crabs, lobsters), and myriapods (e.g., centipedes, millipedes). Each group within the phylum Arthropoda exhibits specific adaptations and characteristics suited to their particular lifestyles and habitats.

Spider

Phylum Mollusca

Phylum Mollusca is a diverse group of invertebrate animals that includes organisms such as snails, clams, octopuses, and squids. Here are some general features of Mollusca:

i. **Soft Body:** Mollusks have soft bodies that are usually covered by a fleshy mantle. The mantle secretes a protective shell in many species, although not all mollusks possess shells.

ii. **Bilateral Symmetry:** Mollusks exhibit bilateral symmetry, meaning their bodies can be divided into two similar halves along a central axis.

iii. **Muscular Foot:** Most mollusks have a muscular foot, which is a specialized structure used for locomotion. The foot may be adapted for crawling, burrowing, or attachment, depending on the species.

iv. **Mantle Cavity:** The mantle cavity is a space between the body and the mantle that houses the gills or lungs. It also serves as an outlet for excretion and reproductive structures.

v. **Radula:** Mollusks typically possess a specialized feeding organ called a radula. The radula consists of tiny, tooth-like structures that scrape or rasp food particles from surfaces.
vi. **Coelom:** Mollusks have a true coelom, which is a fluid-filled body cavity that houses various internal organs.
vii. **Nervous System:** Mollusks have a relatively well-developed nervous system. They possess a pair of ganglia (clusters of nerve cells) and nerve cords that connect to various parts of the body.
viii. **Open Circulatory System:** Most mollusks have an open circulatory system, meaning their blood is not confined to blood vessels. Instead, it flows freely in body cavities and bathes the organs.
ix. **Respiratory Structures:** Mollusks employ various respiratory structures. Some species have gills for extracting oxygen from water, while others have lungs or specialized mantle surfaces for gas exchange.
x. **Reproduction:** Mollusks exhibit a range of reproductive strategies. Most species have separate sexes (dioecious), while some are hermaphroditic, possessing both male and female reproductive organs.

Examples of Mollusca include: Snails (Class Gastropoda), Snails (Class Gastropoda), Squids, octopuses, and cuttlefish (Class Cephalopoda), Chitons (Class Polyplacophora).

Introduction To Biology

©2005 California Academy of Sciences
Achatina immaculata; Giant African Snail

Phylum Echinodermata

Echinodermata is a phylum of marine animals that includes a diverse group of organisms such as starfish, sea urchins, sea cucumbers, and sea lilies.

i. **Radial Symmetry:** Echinoderms typically exhibit radial symmetry, which means their body parts are arranged around a central axis. However, some echinoderms, like sea cucumbers, may have bilateral symmetry during certain stages of their life cycle.

ii. **Spiny Skin:** Echinoderms are characterized by a unique feature called spiny skin. Their skin is often covered with calcareous plates or spines that provide protection and support.

iii. **Water Vascular System:** Echinoderms have a unique hydraulic system called the water vascular system. This system consists of a network of water-filled canals and tube feet that are used for locomotion, respiration, and capturing food.

iv. **Tube Feet:** Tube feet are small, flexible, and hollow structures found on the undersides of echinoderms. These tube feet are controlled by the water vascular system and are used for various functions, including locomotion, feeding, and attachment.

v. **Endoskeleton:** Echinoderms possess an internal skeleton known as an endoskeleton. The endoskeleton is composed of calcareous plates or ossicles that are interconnected and provide structural support for the body.

vi. **Pentaradial Symmetry:** Most echinoderms exhibit pentaradial symmetry, meaning their bodies can be divided into five roughly equal parts radiating from the center. This symmetry is a derived characteristic of echinoderms, as they evolved from bilaterally symmetrical ancestors.

vii. **Regeneration:** Echinoderms have impressive regenerative abilities. They can regenerate lost body parts, such as arms or even entire individuals, given the appropriate conditions.

viii. **Water Vascular System Extensions**: Some echinoderms have modified extensions of their water vascular system. For example, sea stars have tube feet that extend externally from their bodies, while sea urchins have long spines attached to the skeleton through the water vascular system.

ix. **Unique Feeding Mechanisms:** Echinoderms employ various feeding strategies. For example, starfish (sea stars) use their tube feet to pry open bivalve shells, while sea urchins use specialized mouthparts called Aristotle's lantern to scrape algae and other food sources.

x. **Secondary Symmetry:** Some echinoderms, such as sea cucumbers, exhibit secondary bilateral symmetry during certain stages of their life cycle. This allows them to move and interact with their environment in different ways.

These features collectively make echinoderms a distinct and fascinating group of marine organisms. Their diverse adaptations and ecological roles contribute to the overall biodiversity and ecological balance of marine ecosystems.

Introduction To Biology

©2013 Simon J. Tonge
Asterias rubescens; Common Starfish

Phylum Hemichordata

Hemichordata is a phylum of marine invertebrates that are characterized by a combination of features from both chordates (the phylum that includes vertebrates) and echinoderms (the phylum that includes starfish and sea urchins).

i. **Body Structure:** Hemichordates have a body plan consisting of three main parts: the proboscis, collar, and trunk. The proboscis is a muscular extension used for feeding and burrowing. The collar region contains structures involved in filter feeding, gas exchange, and reproduction. The trunk is the elongated main body region.

ii. **Pharyngeal Slits:** Hemichordates possess pharyngeal slits, which are openings in the pharynx (throat) that function in filter feeding and gas exchange. These slits are similar to the gill slits found in chordates and play a role in water circulation and respiration.

iii. **Stomochord:** A unique feature of Hemichordata is the presence of a stomochord, which is a rod-like structure that runs beneath the pharynx. The stomochord provides support and may have a role in gas exchange and feeding.
iv. **Nervous System:** Hemichordates have a relatively simple nervous system consisting of a nerve cord that runs along the underside of the body and a series of ganglia (nerve cell clusters) that control various functions.
v. **Acorn Worms and Pterobranchs:** Hemichordata includes two main groups: the acorn worms (Enteropneusta) and the pterobranchs (Pterobranchia). Acorn worms are typically burrowing marine organisms with elongated bodies, while pterobranchs are colonial filter feeders that live in tubes.

Examples of organisms in the phylum Hemichordata: Acorn Worms (Enteropneusta): Saccoglossus spp., Balanoglossus spp., Ptychodera flava,

Pterobranchs (Pterobranchia): Cephalodiscus spp., Rhabdopleura spp.

These are just a few examples of the organisms found within the phylum Hemichordata. Hemichordates are a relatively small and less well-known group compared to other marine invertebrates, but they play important roles in marine ecosystems and contribute to the overall biodiversity of the ocean.

Pterobranchs

Phylum Chordata

Phylum Chordata encompasses a diverse group of animals that possess certain distinctive features throughout at least some stage of their life cycle. Chordata is a vast phylum that includes several subphyla and diverse organisms. The most well-known subphylum is Vertebrata, which comprises animals with a backbone, such as fish, reptiles, birds, and mammals. However, it's important to note that not all chordates are vertebrates, and there are non-vertebrate chordates like tunicates and lancelets that exhibit these chordate features as well. Here are some features of Phylum Chordata:

i. **Notochord:** Chordates possess a notochord, which is a flexible rod-like structure located along the dorsal (back) side of the body. The notochord provides support and serves as an axis for muscle attachment.

ii. **Dorsal Nerve Cord:** Chordates have a dorsal nerve cord, which is a hollow tubular structure located on the dorsal side of the body. It develops from the ectoderm and becomes the central nervous system, including the brain and spinal cord.

iii. **Pharyngeal Slits or Pouches:** Chordates exhibit pharyngeal slits or pouches, which are openings in the pharynx (throat) region. In some chordates, these structures are involved in filter feeding or respiration. In vertebrates, they develop into various structures such as gills, parts of the ear, or contribute to the formation of the neck and throat.
iv. **Post-Anal Tail:** Chordates possess a tail that extends beyond the anus at some point in their life cycle. This tail may be present during the embryonic stage or persist into adulthood. It provides balance, locomotion, or propelling force in aquatic species.
v. **Endostyle or Thyroid Gland:** Chordates have an endostyle or thyroid gland, which is a glandular structure located in the pharyngeal region. It is involved in the production of various hormones and plays a role in regulating metabolism and development.
vi. **Bilateral Symmetry:** Chordates exhibit bilateral symmetry, meaning their bodies can be divided into two equal halves along a central axis.

Classification Of Phylum Chordata

Phylum Chordata is classified into three subphyla, namely
- Urochordata (tunicates),
- Cephalochordata (lancelets)
- Vertebrata (vertebrates).

The subphylum Urochordata and Cephalochordata are collectively known as protochordates, which are marine animals. They are invertebrates but they share attributes of chordates.

Subphylum Urochordata

Urochordata, also known as tunicates or sea squirts, is a subphylum of Chordata. Although they may not resemble typical chordates in their adult form, they exhibit chordate characteristics during their larval stage. Here are some features of Urochordata:

i. **Tunic:** Urochordates are named after their unique outer covering called the tunic. The tunic is a thick, leathery or gelatinous covering that surrounds the body of the organism. It provides protection and support.

ii. **Filter Feeding:** Urochordates are filter feeders. They possess a specialized structure called the pharyngeal basket or pharyngeal gill slits that are used to filter small particles, such as plankton, from the water. The water is drawn into the body through an incurrent siphon, passes through the pharynx, and exits through an excurrent siphon.

iii. **Incurrent and Excurrent Siphons:** Urochordates have two siphons, an incurrent siphon and an excurrent siphon. These siphons allow water to flow through the body, facilitating filter feeding and respiration.

iv. Notochord and Nerve Cord: Urochordate larvae possess a notochord and a dorsal nerve cord, which are defining features of chordates. However, these structures are typically lost or greatly reduced in the adult stage.

v. **Reduced Body Symmetry:** While the larval stage of urochordates exhibits bilateral symmetry, the adult forms may have reduced symmetry. They often have a cylindrical or sac-like body shape with various protrusions and openings.

vi. **Hermaphroditic Reproduction:** Urochordates are mostly hermaphroditic, meaning they have both male and female reproductive organs. They can reproduce both sexually, by releasing eggs and sperm into the water, and asexually, through budding or fragmentation.

vii. **Larvaceans and Sea Squirts:** Urochordates include two major groups: larvaceans and sea squirts. Larvaceans are planktonic and retain their chordate characteristics throughout their entire life cycle. Sea squirts, on the other hand, undergo a metamorphosis during development, resulting in a loss of chordate features in their adult form.

Urochordates are predominantly marine organisms found in a variety of marine environments worldwide. They play important ecological roles as filter feeders and contribute to nutrient cycling in marine ecosystems.
Examples of organisms in this subphla:
Sea Squirts (Class Ascidiacea): Botryllus schlosseri (Golden star tunicate), Ciona intestinalis (Sea vase), Atriolum robustum
Larvaceans (Class Appendicularia): Oikopleura dioica, Fritillaria borealis
Thaliaceans (Class Thaliacea): Salps, Doliolids

©2005 California Academy of Sciences
Atriolum robustum; Sea Squirts

Phylum Cephalochordata

Cephalochordata, commonly known as lancelets or amphioxus, is a subphylum of Chordata that comprises a small group of marine organisms. Lancelets are small, fish-like creatures that exhibit several notable characteristics. Here are some features of Cephalochordata:

i. **Body Shape and Size:** Lancelets have a slender, elongated body shape resembling a fish. They typically range in size from a few centimeters to a few inches in length.
ii. **Filter Feeding:** Lancelets are filter feeders, using their oral hood and numerous gill slits to capture small particles of food from the water. They draw in water through their mouth and expel it through the gill slits, where food particles are trapped and ingested.
iii. **Notochord and Dorsal Nerve Cord:** Like other chordates, lancelets possess a notochord, which provides support and flexibility. They also have a dorsal nerve cord running along the length of their body, serving as their central nervous system.
iv. **Paired Muscular Segments (Myomeres):** The body of lancelets is divided into a series of similar, segmented blocks known as myomeres. These segments contain skeletal muscles responsible for the undulating swimming motion of lancelets.
v. **Lack of Vertebrae and Cranium:** Unlike vertebrates, lancelets do not possess true vertebrae or a distinct cranium. Their notochord persists throughout their life, providing support and flexibility without the need for bony structures.
vi. **Burrowing Habitats:** Lancelets are primarily burrowers, dwelling in sandy or muddy substrates along shallow coastal areas. They spend much of their time partially buried in the sediment, with only their anterior region exposed.
vii. **Sexually Reproductive:** Lancelets are sexually reproductive organisms, with separate sexes (male and female). They release eggs and sperm into the water during spawning, where fertilization takes place. The fertilized eggs develop

into free-swimming larvae before settling and undergoing metamorphosis into the adult lancelet form.

Examples of organism in this subphyla are: Branchiostoma lanceolatum, Asymmetron lucayanum, Epigonichthys maldivensis.

©2020 Dr. Daniel L. Geiger
Branchiostoma lanceolatum; Lancet Fish

Subphylum Vetebrata

Vertebrata, or vertebrates, is a diverse subphylum of animals within the phylum Chordata. Vertebrates are characterized by the presence of a vertebral column or backbone, which provides structural support and protection for the spinal cord. This subphylum includes a wide range of animals, from fish and amphibians to reptiles, birds, and mammals. The characteristic features of vertebrates include:

These are advanced chordates and have cranium around the brain. The notochord is replaced by a vertebral column in adults. This is why it is said that all vertebrates are chordates but all chordates are not vertebrates:

i. A high degree of cephalization is observed.

ii. The epidermis is multi-layered.
iii. They consist of three types of muscles-striped, unstriped and cardiac.
iv. They have a well-developed coelom.
v. The alimentary canal is complete.
vi. The heart is three or four-chambered.
vii. They have well-developed respiratory and excretory systems.
viii. Endocrine glands are present in all.
ix. They are unisexual and reproduce sexually, hagfish being an exception.

Class Agnatha

Class Agnatha is a group of jawless fish within the subphylum Vertebrata. The name Agnatha is derived from the Greek words "a," meaning without, and "gnathos," meaning jaw. The class includes two main groups of jawless fish: hagfish and lampreys. Characteristics and features of Class Agnatha include:

i. **Jawless:** As the name suggests, Agnathans lack jaws, which distinguishes them from other classes of fish. Instead of jaws, they have circular, toothed mouthparts.
ii. **Cartilaginous Skeleton:** Agnathans have a cartilaginous skeleton, which is flexible and lighter than a bony skeleton. This characteristic allows them to navigate through narrow crevices and burrows.
iii. **Lack of Paired Fins:** Unlike most other fish, Agnathans lack paired fins, such as pectoral and pelvic fins. Instead, they have a long, eel-like body shape that helps them swim and maneuver.
iv. **Notochord Persistence:** Agnathans retain a notochord throughout their entire life cycle. The notochord provides structural support and acts as a precursor to the backbone found in more advanced vertebrates.

v. **External Slime Production:** Both hagfish and lampreys secrete copious amounts of slime. This slime serves various purposes, including defense against predators, reducing friction during movement, and facilitating feeding.
vi. **Parasitic and Scavenging Behavior:** Lampreys are often parasitic, using their toothed, sucker-like mouth to attach to the bodies of other fish and feed on their blood and body fluids. Hagfish, on the other hand, are scavengers, feeding on decaying organisms on the ocean floor.
vii. **Primitive Circulatory System:** Agnathans have a relatively simple circulatory system, lacking a true heart and consisting of a series of vessels that transport blood and other fluids.

Lampreys

Class Chondrichthyes

Class Chondrichthyes, commonly known as cartilaginous fish, is a class of vertebrates within the subphylum Vertebrata. They are characterized by their cartilaginous skeletons, paired fins, and well-developed jaws. Characteristics of Class Chondrichthyes are:

i. **Cartilaginous Skeleton:** Unlike bony fish, which have skeletons made of bone, cartilaginous fish have skeletons made of cartilage. This cartilaginous framework provides flexibility and reduces the overall weight of the fish.
ii. **Paired Fins:** Chondrichthyes have paired pectoral and pelvic fins. These fins help with stability, maneuverability, and steering while swimming.
iii. **Jaws and Teeth:** Cartilaginous fish have well-developed jaws with numerous rows of sharp teeth. Their jaws are not fused to the skull, allowing for a greater range of motion and a more powerful bite. Their teeth are continuously replaced throughout their lives, ensuring efficient feeding and predator-prey interactions.
iv. **Placoid Scales:** Chondrichthyes have unique scales called placoid scales or dermal denticles. These scales are small, tooth-like structures that cover their skin, providing protection and reducing drag in the water.
v. **Internal Fertilization:** Most chondrichthyans exhibit internal fertilization, where the male introduces sperm into the female's body through specialized structures called claspers. This adaptation increases the likelihood of successful reproduction in aquatic environments.
vi. **Osmoregulation:** Cartilaginous fish have specialized organs called rectal glands or salt glands that help them regulate the salt concentration in their bodies. This adaptation allows them to maintain osmotic balance in various salinity environments.
vii. **Marine Habitats:** Chondrichthyans are predominantly marine species, found in a wide range of marine habitats, including oceans, coral reefs, and deep-sea environments. Some species, like bull sharks and sawfish, can tolerate both saltwater and freshwater environments.
viii. **Diverse Feeding Habits:** Cartilaginous fish exhibit a wide range of feeding habits. Some are predators that feed on other

fish, squid, or marine mammals, while others are filter feeders, feeding on plankton and small organisms.

Examples of Chondrichthyans: Class Chondrichthyes includes various well-known species, such as:

Sharks: Great white shark, tiger shark, hammerhead shark, etc.
Rays: Manta ray, stingray, electric ray, etc.
Skates: Common skate, longnose skate, thorny skate, etc.

©2012 Douglas Klug
Carcharodon carcharias; Great White Shark

Class Amphibia

Class Amphibia represents a group of vertebrates that includes amphibians. Amphibians are unique organisms that exhibit a combination of aquatic and terrestrial adaptations throughout their life cycle. Features of Class Amphibia include:

i. **Skin and Glandular Secretions:** Amphibians have thin, moist, and permeable skin that allows them to exchange gases, such as oxygen and carbon dioxide, with the environment. Their skin is often glandular and produces mucus, which helps keep it moist and aids in respiration. Some amphibians also have specialized skin glands that secrete toxins or chemicals used for defense or communication.

ii. **Metamorphosis:** Amphibians typically undergo metamorphosis, a transformation from an aquatic larval form to a semi-aquatic or terrestrial adult form. For example, frogs and toads start as aquatic tadpoles with gills, which eventually undergo metamorphosis to develop lungs and limbs, enabling them to live on land.

iii. **Dual Respiratory System:** Amphibians have a dual respiratory system that allows them to respire both in water and on land. While aquatic larvae primarily respire through gills, adult amphibians use lungs for respiration. However, they can also absorb oxygen through their thin and moist skin.

iv. **Ectothermic Regulation:** Amphibians are ectothermic, meaning they rely on external sources of heat to regulate their body temperature. They are typically more active and efficient at higher temperatures, as their metabolism depends on environmental warmth.

v. **Aquatic and Terrestrial Lifestyles:** Amphibians have adapted to a wide range of habitats. Many species have an aquatic larval stage, living in freshwater bodies such as ponds or streams, before transitioning to a semi-aquatic or terrestrial adult form. Some amphibians, such as salamanders, remain fully aquatic throughout their lives, while others, like frogs and toads, are predominantly terrestrial but require access to water for reproduction.

vi. **Moist Environment Dependence:** Amphibians rely on a moist environment for various aspects of their life, including reproduction, respiration, and maintaining hydration. Their permeable skin necessitates moisture to prevent dehydration.

vii. **Dual Circulatory System:** Amphibians possess a three-chambered heart that pumps oxygenated blood to body tissues and deoxygenated blood to the lungs or skin for oxygen exchange. Their circulatory system allows for some separation of oxygenated and deoxygenated blood but is less efficient compared to the four-chambered heart of reptiles, birds, and mammals.

Examples of Amphibians: Class Amphibia includes various well-known amphibian groups and species, such as: Frogs, Toads, Salamanders, axolotl, newt, hellbender, Caecilians: Legless amphibians that resemble large worms or snakes, found in tropical regions.

©1979 Alan Resetar
Acris blanchardi; Blanchard's Cricket Frog

Class Reptilia

Class Reptilia represents a diverse group of vertebrates known as reptiles. Reptiles have a number of distinctive features and adaptations that set them apart from other animal classes. Here are some key features of Class Reptilia:

i. **Dry and Scaly Skin:** Reptiles have dry and scaly skin, which provides protection against water loss and external injuries.

Their scales are made of keratin and serve as a barrier to the environment.

ii. **Ectothermic Regulation:** Reptiles are ectothermic, meaning they rely on external sources of heat to regulate their body temperature. They are often referred to as "cold-blooded" because their body temperature fluctuates with the surrounding environment.

iii. **Amniotic Eggs:** Reptiles lay amniotic eggs with a protective shell, which allows them to reproduce on land. The amniotic eggs contain specialized membranes that provide nourishment, waste disposal, and gas exchange, enabling the embryo to develop outside of water.

iv. **Lungs for Respiration:** Reptiles breathe air using lungs. Unlike amphibians, reptiles lack gills and rely solely on lungs for respiration throughout their life cycle.

v. **Strong and Efficient Heart:** Reptiles have a three-chambered heart (except for some crocodilians, which have four-chambered hearts), which separates oxygenated and deoxygenated blood to some extent. Although less efficient than the four-chambered hearts of birds and mammals, reptilian hearts enable improved oxygenation of tissues compared to amphibians.

vi. **Terrestrial Adaptations:** Most reptiles are well-adapted for terrestrial life. Their limbs are positioned in a sprawled or erect posture, allowing for efficient movement on land. Some reptiles, such as snakes, have elongated bodies and lack limbs entirely.

vii. **Scales and Shells:** Reptiles have a variety of external coverings. Some have scales, while others have shells. Turtles, for example, possess a protective shell composed of bony plates fused to their skeleton.

viii. **Carnivorous or Herbivorous Diet:** Reptiles exhibit diverse feeding habits. Some are carnivorous, preying on other animals, while others are herbivorous, feeding primarily on plants. This diversity is evident in reptilian orders such as Crocodylia, Squamata (which includes lizards and snakes), and Testudines (turtles and tortoises).

Examples of Reptiles: Class Reptilia includes various well-known reptile groups and species, such as: Snakes, Lizards, Turtles and Tortoises, Crocodilians

©2001 California Academy of Sciences
Crocodylus acutus; American Crocodile

Class Aves

Class Aves represents the group of animals commonly known as birds. Birds are highly specialized and diverse creatures, adapted for flight and occupying a wide range of habitats. Here are some key features of Class Aves:

i. **Feathers:** Feathers are one of the defining characteristics of birds. They provide insulation, enable flight, and assist in courtship displays. Feathers are composed of keratin and are constantly maintained through preening.

ii. **Wings and Flight:** Birds have modified forelimbs known as wings that enable them to fly. The structure and shape of wings vary among bird species, reflecting their different flight capabilities and habits. Flight allows birds to travel efficiently, find food, escape predators, and migrate long distances.

iii. **Lightweight Skeleton:** Birds have a lightweight skeleton, with hollow bones filled with air cavities. This adaptation reduces overall weight, making flight easier. Some bones in the skeleton are fused for increased strength and stability during flight.

iv. **Beak and Jaw Adaptations:** Birds have a beak instead of teeth. The shape and size of the beak are adapted to their specific feeding habits. Beaks can be sharp and pointed for catching prey, stout for cracking seeds and nuts, or long and curved for probing flowers or extracting nectar.

v. **Endothermic Regulation:** Birds are endothermic, meaning they generate and maintain their body temperature through internal heat production. This allows birds to be active and maintain high metabolic rates, even in diverse environments and climates.

vi. **Efficient Respiratory System:** Birds have a highly efficient respiratory system that allows for efficient oxygen uptake. Air sacs connected to their lungs help maintain a continuous flow of fresh oxygenated air, allowing for high metabolic rates needed for flight.

vii. **Four-Chambered Heart:** Birds have a four-chambered heart, which allows for complete separation of oxygenated and deoxygenated blood. This enables efficient oxygenation of tissues and supports the high energy demands of flight.

viii. **Efficient Digestive System:** Birds have a specialized digestive system adapted to their varied diets. Their digestive tract includes crop and gizzard structures that aid in food storage, grinding, and digestion of tough materials.

ix. **Reproduction and Care of Young:** Most birds lay hard-shelled eggs and exhibit diverse nesting behaviors. Many species engage in elaborate courtship displays and build intricate nests for egg-laying. Birds also invest significant time and effort in caring for their young, providing food and protection until they can fend for themselves.

x. **Vocalization and Communication:** Birds are known for their diverse and complex vocalizations, used for communication, mate attraction, and territorial defense. They have specialized vocal organs called syrinx, located at the base of their trachea, which enable the production of a wide range of sounds.

Examples of birds: Class Aves encompasses a vast array of bird species, including:
- **Raptors:** Eagles, hawks, falcons, owls.
- **Passerines:** Sparrows, finches, warblers, crows, robins
- **Waterbirds:** Ducks, geese, swans, herons, flamingos
- **Flightless Birds:** Ostriches, emus, penguins, kiwis

©2013 California Academy of Sciences
Asio flammeus galapagoensis; Short-eared Owls

Class Mammalia

Class Mammalia represents a diverse group of animals known as mammals. Mammals are characterized by several distinctive features that set them apart from other animal classes. Here are some key features of Class Mammalia:

i. **Mammary Glands and Milk Production:** Mammals possess mammary glands that produce milk, which is used to nourish their offspring. Milk provides essential nutrients and antibodies, promoting the growth and development of young mammals.

ii. **Hair or Fur:** Mammals have hair or fur covering their bodies. Hair provides insulation, protection, and sensory functions. It varies in texture, color, and density among different mammalian species.

iii. **Endothermic Regulation:** Mammals are endothermic, meaning they regulate their body temperature internally. This allows them to maintain a relatively constant body temperature regardless of environmental conditions.

iv. **Specialized Teeth:** Mammals have different types of teeth adapted to their specific diets. These include incisors for biting, canines for tearing, premolars for grinding, and molars for chewing. The number and arrangement of teeth vary among different mammalian groups.

v. **Well-Developed Brain:** Mammals have a well-developed brain that allows for complex cognitive abilities, learning, and behavioral adaptations. They exhibit diverse behaviors, including social interactions, communication, problem-solving, and tool use.

vi. **Diaphragm and Efficient Respiration:** Mammals possess a diaphragm, a muscle located beneath the lungs that aids in respiration. This allows for more efficient breathing and increased lung capacity.

vii. **Four-Chambered Heart:** Mammals have a four-chambered heart, which ensures efficient separation of oxygenated and deoxygenated blood. This allows for efficient oxygen delivery to body tissues.
viii. **Viviparity:** Most mammals give birth to live young, a reproductive strategy known as viviparity. The development of embryos occurs internally, and offspring are nourished through the placenta before birth.
ix. **Diverse Reproductive Strategies:** Mammals exhibit a range of reproductive strategies, including monogamy, polygamy, and various mating systems. Some species have prolonged parental care and complex social structures.
x. **Specialized Limbs:** Mammals have limbs adapted for diverse modes of locomotion. This includes running, climbing, swimming, digging, and flying. Limb structures vary greatly among different mammalian groups to suit their specific habitats and behaviors.
xi. **Extensive Brain Development:** Mammals have a well-developed cerebral cortex, which is responsible for higher cognitive functions, memory, and sensory perception. This contributes to their ability to adapt to various environments and exhibit complex behaviors.

Examples of Mammals: Class Mammalia encompasses a wide range of mammalian species, including:
- ✓ **Primates:** Humans, monkeys, apes
- ✓ **Carnivores:** Lions, tigers, wolves, bears
- ✓ **Herbivores:** Cows, deer, elephants, horses
- ✓ **Rodents:** Mice, rats, squirrels
- ✓ **Cetaceans:** Dolphins, whales, porpoises

Introduction To Biology

©2000 John White
Panthera Leo; Lion

■ Plant Kingdom

The Plant Kingdom, also known as Kingdom Plantae, encompasses a wide variety of multicellular organisms that are capable of photosynthesis. Plants play a crucial role in the Earth's ecosystems, serving as primary producers and providing oxygen, food, shelter, and numerous other benefits to both humans and other organisms.

i. **Eukaryotic Cells:** Plants are composed of eukaryotic cells, meaning they have a true nucleus and membrane-bound organelles.

ii. **Autotrophic Nutrition:** Plants are autotrophs, capable of synthesizing their own food through the process of photosynthesis. They use chlorophyll, a pigment found in chloroplasts, to capture sunlight and convert it into chemical energy in the form of glucose.

iii. **Cell Wall:** Plant cells have a rigid cell wall composed of cellulose, which provides structural support and protection for the plant.
iv. **Multicellularity:** Plants are multicellular organisms, meaning they are composed of multiple cells that work together to form tissues, organs, and organ systems.
v. **Alternation of Generations:** Plants exhibit a life cycle that involves alternation between a multicellular haploid gametophyte generation and a multicellular diploid sporophyte generation.
vi. **Reproduction:** Plants have various reproductive strategies. Some reproduce sexually, with the production of gametes (sperm and eggs) that combine to form a zygote. Others can reproduce asexually through methods such as vegetative propagation, where new plants are generated from existing plant parts.
vii. **Specialized Tissues and Organs:** Plants have specialized tissues and organs that enable them to carry out essential functions. These include roots for absorption of water and nutrients, stems for support and transport, leaves for photosynthesis, and flowers for reproduction.
viii. **Vascular System:** Many plants have a vascular system that consists of xylem and phloem. Xylem transports water and minerals from the roots to other parts of the plant, while phloem transports sugars and other organic compounds throughout the plant.
ix. **Adaptations to Land:** Plants have evolved a variety of adaptations to thrive on land. These include the development of cuticles, which reduce water loss through leaves, and the evolution of stomata, small openings on the surface of leaves that regulate gas exchange.
x. **Diverse Forms and Habitats:** The Plant Kingdom exhibits tremendous diversity in terms of form, size, and habitat.

Introduction To Biology

Plants range from tiny mosses and algae to towering trees. They can be found in a wide array of environments, including forests, grasslands, deserts, wetlands, and aquatic ecosystems.

Classification Of Kingdom Plantae

A plant kingdom is further classified into subgroups. Classification is based on the following criteria:
- ✓ **Plant body:** Presence or absence of a well-differentiated plant body. E.g. Root, Stem and Leaves.
- ✓ **Vascular system:** Presence or absence of a vascular system for the transportation of water and other substances. E.g. Phloem and Xylem.
- ✓ **Seed formation:** Presence or absence of flowers and seeds and if the seeds are naked or enclosed in a fruit.

The plant kingdom has been classified into five subgroups according to the above-mentioned criteria:
- ➢ Thallophyta
- ➢ Bryophyta
- ➢ Pteridophyta
- ➢ Gymnosperms
- ➢ Angiosperms

Thallophyta

Thallophyta is a division within the Plant Kingdom that includes a diverse group of non-vascular plants. These plants lack true roots, stems, and leaves, and instead have a thallus, which is a simple, undifferentiated body. Thallophyta is also known as the "thalloid plants" or "lower plants." Characteristics of thallopyte:
i. **Body Structure:** Thallophyta plants have a simple and flattened body structure known as a thallus. The thallus is not

organized into distinct organs like roots, stems, or leaves. Instead, it is a flattened or branching structure that performs various functions such as photosynthesis, absorption, and reproduction.

ii. **Lack of Vascular Tissue:** Thallophyta plants lack specialized vascular tissues, including xylem and phloem. As a result, they do not have true roots, stems, or leaves for the transport of water, nutrients, and sugars. Instead, they rely on direct diffusion or osmosis for nutrient uptake.

iii. **Reproduction:** Thallophyta plants reproduce through various methods. They can reproduce sexually, producing gametes that combine to form a zygote. They can also reproduce asexually through methods such as fragmentation, where a part of the thallus breaks off and develops into a new individual.

iv. **Habitat and Diversity:** Thallophyta plants can be found in a wide range of habitats, including terrestrial, freshwater, and marine environments. They are often found in moist or aquatic habitats, where they can absorb water and nutrients directly through their thallus. Thallophyta includes various groups of organisms such as algae, lichens, and certain types of fungi.

v. **Algae:** Algae are a prominent group within Thallophyta. They are photosynthetic organisms that can be unicellular, colonial, or multicellular. Algae can be classified into different divisions based on their pigmentation, cell wall composition, and mode of reproduction. Examples of algae include green algae, red algae, brown algae, and diatoms.

vi. **Lichens:** Lichens are symbiotic associations between fungi and algae or cyanobacteria. The fungal partner provides a protective structure, while the algal or cyanobacterial partner carries out photosynthesis. Lichens are often found in harsh environments and play a crucial role in soil formation and ecosystem stability.

vii. **Importance:** Thallophyta plants play important ecological roles. They are primary producers, converting sunlight into organic matter through photosynthesis. They also contribute to oxygen production, carbon dioxide fixation, and nutrient cycling in various ecosystems. Thallophyta plants are also used by humans for various purposes, such as food (edible seaweeds), medicines, and industrial applications.

Thallophyta represents a diverse and fascinating group of plants that have unique adaptations and contribute significantly to the ecological balance of different habitats. While they may be simpler in structure compared to higher plants, they are essential components of ecosystems and contribute to the overall biodiversity of the Plant Kingdom.

Bryophyta

Bryophyta is a division within the Plant Kingdom that includes the group of plants commonly known as mosses. These plants are small, non-vascular, and usually found in moist environments. Bryophyta is also referred to as the "bryophytes" or "moss plants."

i. **Non-Vascular Structure:** Bryophyta plants lack true vascular tissues, including xylem and phloem. As a result, they do not have specialized structures for water and nutrient transport, such as roots, stems, or leaves. Instead, they absorb water and nutrients directly through their tissues.
ii. **Simple Body Structure:** Bryophyta plants have a relatively simple body structure. They consist of leaf-like structures called "leaves," which are arranged in a spiral or overlapping pattern around a stem-like structure called the "stem." The stem and leaves are typically only a few cells thick.
iii. **Rhizoids:** Bryophyta plants have hair-like structures called "rhizoids" that anchor them to the substrate. Rhizoids also help in the absorption of water and nutrients from the surrounding environment.
iv. **Dominant Gametophyte Generation:** The life cycle of Bryophyta plants is dominated by the gametophyte generation, which is the haploid phase of the plant. The gametophyte is the sexual phase that produces gametes (sperm and eggs) through specialized structures called gametangia.
v. **Reproduction:** Bryophyta plants reproduce through alternation of generations, involving both sexual and asexual modes of reproduction. The gametophyte generation produces male and female gametangia. Sperm produced in the antheridia fertilize the eggs produced in the archegonia, leading to the formation of a diploid zygote. The zygote develops into a sporophyte, which remains attached to the gametophyte and produces spores through meiosis. The spores are released and germinate into new gametophytes.
vi. **Habitat and Diversity:** Bryophyta plants are commonly found in moist environments, including forests, wetlands, and areas

with high humidity. They can tolerate shade and are often seen covering rocks, soil, and tree trunks. Bryophyta encompasses a diverse group of plants, including various species of mosses.

vii. **Ecological Importance:** Bryophyta plants play important ecological roles. They help in soil formation and contribute to water retention in terrestrial ecosystems. They provide habitats for small invertebrates, serve as food for certain animals, and contribute to the overall biodiversity of ecosystems.

viii. **Environmental Indicators:** Certain species of mosses within Bryophyta are used as environmental indicators. Their presence or absence in specific habitats can provide valuable information about environmental conditions, such as air pollution levels, moisture availability, and habitat quality.

The main example of the organism in this group is mosses

Mosses

Pteridophyta

Pteridophyta is a division within the Plant Kingdom that includes the group of plants commonly known as ferns and fern allies.

These plants are vascular and reproduce through spores. Pteridophytes have a long evolutionary history and were dominant in the Earth's ecosystems during the Carboniferous period. Characteristics of Pteridophyta:

i. **Vascular Tissues:** Pteridophytes have well-developed vascular tissues, including xylem and phloem, which allow for efficient transport of water, nutrients, and sugars throughout the plant. This vascular system provides support and allows for the growth of larger plant structures.

ii. **Leaves:** Pteridophytes have leaves known as fronds. Fronds are usually large, with divided leaflets called pinnae. The leaves are often coiled when they first emerge, gradually uncoiling as they mature.

iii. **Sporophyte Dominance:** The life cycle of pteridophytes is dominated by the sporophyte generation, which is the diploid phase of the plant. The sporophyte produces spores in structures called sporangia, typically located on the undersides of the fronds.

iv. **Reproduction:** Pteridophytes reproduce through alternation of generations, involving both sexual and asexual modes of reproduction. The sporophyte produces spores through meiosis. These spores are released into the environment, where they can germinate and develop into gametophytes. The gametophytes produce gametes (sperm and eggs), which combine during fertilization to form a zygote. The zygote develops into a new sporophyte.

v. **Habitat and Diversity:** Pteridophytes are found in a wide range of habitats, from moist and shady environments to open areas. They are particularly abundant in tropical rainforests and temperate regions. Pteridophyta includes a diverse group of plants, with ferns being the most well-known and numerous. Other examples include horsetails (Equisetum) and club mosses (Lycopodium).

vi. **Ecological Importance:** Pteridophytes play important ecological roles. They contribute to soil formation, help stabilize slopes, and provide habitats for various organisms. Ferns, in particular, are often used in landscaping and gardening due to their ornamental value and ability to thrive in shaded areas.

vii. **Evolutionary Significance:** Pteridophytes played a significant role in the evolution of plants on Earth. They were among the first plants to possess vascular tissues, allowing for increased size and complexity. They were also important contributors to the formation of coal during the Carboniferous period, as their remains became fossilized and transformed into coal deposits.

©2013 Eric White
Adenophorus hymenophylloides; Palai Huna (hidden Fern)

Gymnosperm

Introduction To Biology

Gymnosperms are a group of plants that belong to the division Pinophyta (also known as Coniferophyta) within the Plant Kingdom. They are characterized by the production of naked seeds, meaning that the seeds are not enclosed within a fruit. Gymnosperms are a diverse group of plants that include conifers, cycads, Ginkgo biloba, and gnetophytes. Features of gymnosperms include:

i. **Reproduction:** Gymnosperms have reproductive structures called cones. They produce male and female cones, also known as strobili. Male cones produce pollen grains that contain the male gametes (sperm), while female cones contain ovules that contain the female gametes (eggs). The transfer of pollen to the female cones is usually achieved through wind pollination.

ii. **Naked Seeds:** One of the defining features of gymnosperms is the production of naked seeds. Unlike angiosperms (flowering plants), which have seeds enclosed within fruits, gymnosperm seeds are not protected by an ovary. Instead, they are directly exposed on the surface of the cone scales or in specialized structures within the cones.

iii. **Vascular Tissues:** Gymnosperms have well-developed vascular tissues, including xylem and phloem, which allow for the efficient transport of water, nutrients, and sugars throughout the plant. This vascular system enables gymnosperms to grow tall and produce woody stems.

iv. **Leaves:** Gymnosperms have needle-like or scale-like leaves, which help reduce water loss through transpiration. These leaves are often evergreen, meaning they remain on the plant year-round.

v. **Adaptations to Harsh Environments:** Gymnosperms have evolved adaptations that allow them to thrive in a variety of environments, including cold and dry regions. They are well-suited to cold climates and can withstand low temperatures, as evidenced by their occurrence in boreal forests and mountainous regions. Additionally, their needle-like leaves and thick cuticles help reduce water loss, enabling them to survive in arid conditions.

vi. **Ecological Importance:** Gymnosperms play important ecological roles. They are often dominant in coniferous forests, where they provide habitats for various organisms. Their needles and cones can serve as food sources for wildlife, and their dense canopies provide shelter and nesting sites for birds and small mammals. Gymnosperms also contribute to carbon sequestration and help regulate the Earth's climate.

vii. **Economic Importance:** Gymnosperms have significant economic value. They are used extensively for timber production, providing lumber for construction, furniture, and paper production. Some gymnosperm species, such as pine and spruce, are also used as Christmas trees. Additionally, certain gymnosperms, such as the Ginkgo biloba tree, have medicinal properties and are used in traditional medicine.

Gymnosperms are a diverse and ancient group of plants that have adapted to various ecological niches. Their unique reproductive strategies, naked seeds, and adaptations to harsh environments make them distinctive within the Plant Kingdom. They contribute to the overall biodiversity of the Earth's ecosystems and have both ecological and economic significance.

Introduction To Biology

©2020 Jean Pawek
Ginkgo biloba

Angiosperm

Angiosperms, also known as flowering plants, represent the largest and most diverse group within the Plant Kingdom. They are characterized by the presence of flowers and the production of seeds enclosed within fruits. Angiosperms have evolved a range of adaptations that have allowed them to dominate terrestrial ecosystems and form complex relationships with animals. Here are some key features and characteristics of angiosperms:

i. **Flowers:** One of the defining features of angiosperms is the presence of flowers. Flowers are reproductive structures that contain male and female reproductive organs. They are typically composed of sepals, petals, stamens (male reproductive organs), and carpels (female reproductive organs). Flowers attract pollinators, such as insects, birds, and bats, which facilitate the transfer of pollen and promote fertilization.

ii. **Fruits and Seeds:** Angiosperms produce seeds that are enclosed within fruits. Fruits are derived from the ovary of the flower and serve to protect and disperse the seeds. Fruits exhibit a wide range of forms, including fleshy fruits (e.g., apples, berries) and dry fruits (e.g., nuts, capsules). The development of fruits and the dispersal of seeds contribute to the reproductive success and dispersal of angiosperms.

iii. **Vascular Tissues:** Angiosperms possess well-developed vascular tissues, including xylem and phloem. Xylem transports water and minerals from the roots to the rest of the plant, while phloem transports sugars and other organic compounds throughout the plant. The presence of vascular tissues allows angiosperms to grow tall and develop complex structures, such as stems, leaves, and flowers.

iv. **Diversity:** Angiosperms exhibit remarkable diversity, with over 300,000 known species. They can be found in almost every habitat on Earth, from rainforests to deserts, and from aquatic environments to high altitudes. This diversity is attributed to their ability to adapt to different ecological conditions, which has led to the evolution of a wide range of plant forms, sizes, and specialized structures.

v. **Pollination:** Angiosperms employ various mechanisms of pollination, including wind pollination and animal pollination. Animal-pollinated flowers often exhibit adaptations to attract specific pollinators, such as bright colors, fragrances, nectar rewards, and specific flower shapes. This mutualistic relationship between angiosperms and pollinators is essential for successful reproduction and the maintenance of plant diversity.

vi. **Ecological Importance:** Angiosperms play crucial ecological roles. They are primary producers, converting sunlight into chemical energy through photosynthesis. They contribute to the oxygen production, carbon sequestration, and nutrient cycling in ecosystems. Angiosperms also provide habitats, food sources, and nesting sites for numerous animals, and they form the basis of many terrestrial food chains.

vii. **Economic Significance:** Angiosperms have immense economic importance. They provide humans with food crops (e.g., grains, fruits, vegetables), spices, fibers (e.g., cotton, flax), medicinal plants, and ornamental plants. Many economic activities, such as agriculture, horticulture, and forestry, rely heavily on angiosperm plants for livelihoods and the global economy.

Angiosperm are split into two classes: Monocotyledonae and Dicotyledonae.

Monocotyledonae (Monocots):

- Monocots have seeds with a single cotyledon (embryonic leaf).
- Their leaves typically have parallel veins.
- Flowers of monocots usually have floral parts in multiples of three.
- Monocots often have fibrous root systems.
- Examples of monocots include grasses (e.g., rice, wheat), lilies, orchids, palms, and bananas.

Introduction To Biology

©2004 California Academy of Sciences
Musa sp.; banana

Dicotyledonae (Dicots):
- Dicots have seeds with two cotyledons.
- Their leaves usually have net-like or branching veins.
- Flowers of dicots typically have floral parts in multiples of four or five.
- Dicots often have taproots or a combination of taproots and fibrous roots.
- Examples of dicots include roses, sunflowers, beans, tomatoes, roses, and oak trees.

2002 Brent Miller
Lycopersicon esculentum; Tomato

Cryptogams and Phanerogams

The plant kingdom is also classified into two groups:
Cryptogams – Non-flowering and non-seed bearing plants. E.g. Thallophyta, Bryophyta, Pteridophyta
Phanerogams – Flowering and seed-bearing plants. E.g. Gymnosperms, Angiosperms

■ Fungi, Protists, And Bacteria

Fungi

The Kingdom Fungi comprises a diverse group of eukaryotic organisms that play crucial roles in various ecosystems. Fungi exhibit distinct characteristics that set them apart from other kingdoms of life.

They are mostly multicellular organisms, although some are unicellular, such as yeasts. They have cell walls composed of chitin, a complex polysaccharide that provides rigidity and support. Fungal cells are organized into thread-like structures called hyphae, which can form extensive networks called mycelium. Hyphae are typically divided into compartments by septa, which contain pores for the movement of nutrients and organelles.

Fungi are heterotrophs, meaning they obtain their nutrients by absorbing organic matter from their surroundings. They secrete enzymes that break down complex organic compounds into simpler molecules, which are then absorbed by the hyphae. Fungi exhibit diverse nutritional strategies, including saprophytic fungi that decompose dead organic material, parasitic fungi that derive nutrients from living hosts, and mutualistic fungi that form symbiotic relationships with other organisms, such as mycorrhizal fungi associating with plant roots.

Fungi can reproduce both sexually and asexually, allowing for genetic variation and adaptation. Sexual reproduction involves the fusion of specialized cells from two different individuals, resulting in the formation of spores with genetic diversity. Asexual reproduction occurs through the production of spores by a single individual, without the need for genetic recombination. Fungal spores can be dispersed by various means, such as wind, water, or through interactions with animals.

Ecological Roles And Importance

1. Fungi play vital roles in ecosystems as decomposers, breaking down dead organic matter and recycling nutrients back into the environment.
2. They contribute to the formation and maintenance of soil by breaking down complex organic compounds into simpler forms.

3. Fungi also form mutualistic relationships with other organisms, such as mycorrhizal associations with plant roots, aiding in nutrient uptake and enhancing plant growth.
4. Some fungi have symbiotic relationships with certain animals, such as lichens formed by the association of fungi and algae or cyanobacteria.
5. Fungi are also involved in the production of various commercial products, including food and beverages (e.g., bread, cheese, beer, wine), antibiotics, enzymes, and biofuels.

Diversity and Examples:

The Kingdom Fungi is incredibly diverse, with an estimated 144,000 known species and potentially many more undiscovered. Fungi can be found in various habitats, including terrestrial, freshwater, and marine environments. Examples of fungi include mushrooms, molds, yeasts, lichens, truffles, and more.

Introduction To Biology

Mushroom

◼ Kingdom Protista

The Kingdom Protista, also known as the protists, is a diverse group of eukaryotic microorganisms that do not fit into the categories of plants, animals, or fungi. They exhibit a wide range of forms, lifestyles, and nutritional strategies, making them a highly varied and fascinating group.

The Kingdom Protista is a polyphyletic group, meaning it includes organisms that do not share a recent common ancestor. Protists are classified into several major groups based on their characteristics and evolutionary relationships. Traditional classification divides protists into three main groups: **animal-like** protists (protozoa), **plant-like protists (algae)**, and **fungus-like protists**. However, due to their diverse nature, the classification of protists is continuously evolving as more is learned about their genetic relationships and ecological roles.

Protists exhibit tremendous diversity in terms of size, complexity, and ecological roles. They can be unicellular (such as amoebas, paramecia, and diatoms) or multicellular (such as seaweeds and slime molds). Protists display a variety of forms, including flagellated, ciliated, and amoeboid forms. They inhabit various habitats, including freshwater, marine environments, moist soils, and even the bodies of other organisms. Protists possess a range of nutritional strategies. Some are autotrophs, capable of photosynthesis to produce their food (e.g., algae). Others are heterotrophs, obtaining nutrients by consuming other organisms (e.g., protozoa). Some protists exhibit mixotrophic capabilities, combining autotrophic and heterotrophic modes of nutrition.

Ecological Significance:

1. Protists play important ecological roles in various ecosystems.
2. Many photosynthetic protists, such as algae, are primary producers and contribute significantly to global oxygen production and carbon fixation.
3. Protists form the base of aquatic food chains, serving as a food source for other organisms, including larger animals like fish and whales.
4. Some protists engage in symbiotic relationships with other organisms. For example, certain protists live within the intestines of termites and aid in the digestion of cellulose.
5. Protists are crucial in nutrient cycling, decomposition, and the recycling of organic matter in ecosystems.

Human Relevance

Introduction To Biology

- Some protists can cause diseases in humans and other animals. For instance, the protozoan Plasmodium causes malaria, while Giardia causes gastrointestinal infections.
- Certain protists have economic importance. For example, certain types of algae are used in food production, such as nori (used in sushi) and spirulina (a nutritional supplement).
- Protists are also studied extensively in scientific research to understand their biology, evolution, and their potential applications in biotechnology.

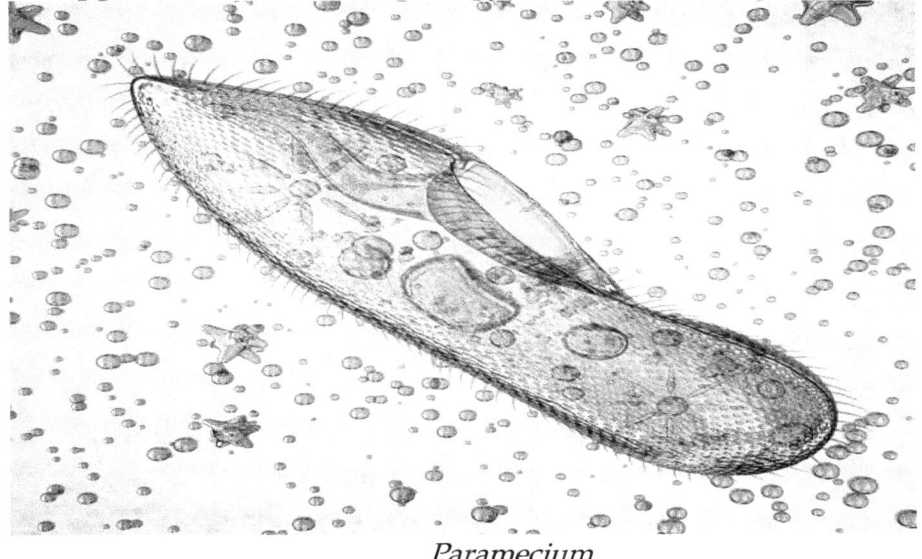

Paramecium

Bacteria

Bacteria are single-celled microorganisms that belong to the domain Bacteria, one of the three domains of life. They are among the most abundant and diverse organisms on Earth and can be found in almost every habitat, including soil, water, air, and even the human body. Bacteria exhibit a wide range of shapes, metabolic capabilities, and ecological roles.

Bacterial cells are prokaryotic, which means they lack a true nucleus and membrane-bound organelles. They have a cell membrane that surrounds the cytoplasm, where essential cellular processes occur. Bacterial cells have a cell wall made of peptidoglycan, a unique compound not found in other organisms. Some bacteria have an additional outer protective layer called the capsule, which provides resistance against environmental stressors and immune system defenses. Bacterial DNA is organized in a circular chromosome located in the nucleoid region of the cell.

Bacteria exhibit a wide range of shapes, including spheres (cocci), rods (bacilli), spirals (spirilla), and filamentous forms. They can occur as single cells, pairs (diplo), chains (strepto), clusters (staphylo), or other arrangements. These different shapes and arrangements can aid in bacterial identification and classification.

Furthermore, bacteria have diverse metabolic capabilities and can be classified based on their energy and carbon sources.

- Autotrophic bacteria obtain energy from inorganic sources and can perform photosynthesis (using light) or chemosynthesis (using inorganic chemicals).
- Heterotrophic bacteria obtain energy by consuming organic compounds produced by other organisms.
- Some bacteria are mixotrophs, capable of both autotrophic and heterotrophic nutrition.

Bacteria exhibit various metabolic pathways, such as fermentation, respiration, and nitrogen fixation.

Ecological Roles

1. Bacteria play vital roles in ecosystems as decomposers, breaking down dead organic matter and recycling nutrients back into the environment.
2. Some bacteria form mutualistic or symbiotic relationships with other organisms, benefiting both parties. Examples include nitrogen-fixing bacteria that form associations with leguminous plants.
3. Bacteria are crucial for nutrient cycling and maintaining ecological balance.
4. Certain bacteria are pathogenic and can cause diseases in humans, animals, and plants. However, it's important to note that only a small fraction of bacteria are harmful, and most bacteria are beneficial or neutral.

Economic Importance

1. Bacteria have immense economic importance and are used in various industrial processes.
2. They are involved in the production of antibiotics, enzymes, vitamins, and other biotechnological products.
3. Bacteria are used in the production of fermented foods, such as yogurt, cheese, and sauerkraut.
4. They are utilized in wastewater treatment, bioremediation, and the production of biofuels.
5. Bacteria also play a significant role in the field of genetic engineering and biotechnology, serving as hosts for gene manipulation and recombinant protein production.

Human Health

Bacteria have a significant impact on human health. While some bacteria can cause diseases, many are essential for maintaining a healthy body. The human microbiome, composed of trillions of bacteria, plays a crucial role in digestion, immune system development, and protection against pathogens. Probiotics, which are live bacteria or bacterial products, are used to promote gut health and improve overall well-being. Bacteria are also used in the production of vaccines and the development of diagnostic tools for infectious diseases.

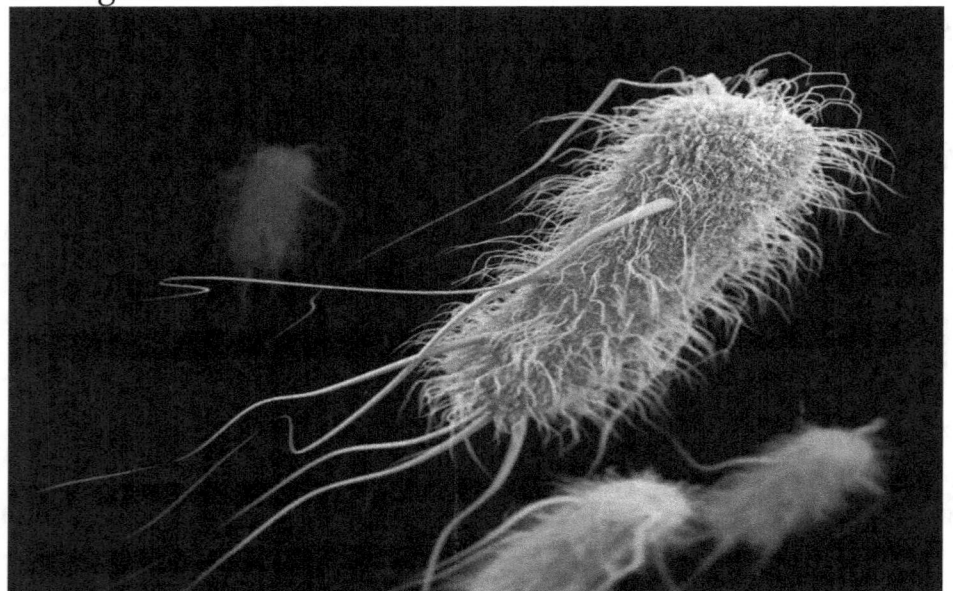

Bacteria

HUMAN ANATOMY AND PHYSIOLOGY — CHAPTER 6

■ Introduction To The Human Body

The human body is an incredibly complex and sophisticated organism composed of various systems, organs, tissues, and cells working together to maintain life. It is a marvel of biological engineering, capable of carrying out numerous functions and processes necessary for survival.

The human body is organized into hierarchical levels, starting from the smallest unit, the cell, and progressing to tissues, organs, and systems. Cells are the building blocks of life and are specialized to perform specific functions within the body. Groups of similar cells form tissues, which further organize to create organs. Organs work together to form systems, and these systems collaborate to maintain the overall function and homeostasis of the body.

Major Systems of the Human Body

The human body consists of several interdependent systems, each responsible for specific functions.

i. The skeletal system provides support, protection, and movement, composed of bones, cartilage, and ligaments.
ii. The muscular system enables movement and generates force through the contraction of muscles.

iii. The circulatory system (cardiovascular system) transports oxygen, nutrients, hormones, and waste products throughout the body using the heart and blood vessels.
iv. The respiratory system facilitates gas exchange, allowing the intake of oxygen and the removal of carbon dioxide, primarily through the lungs.
v. The digestive system processes food, extracting nutrients and eliminating waste, involving organs such as the stomach, intestines, and liver.
vi. The nervous system coordinates body activities and processes sensory information, including the brain, spinal cord, and peripheral nerves.
vii. The endocrine system regulates bodily functions through the production and release of hormones by glands such as the pituitary, thyroid, and adrenal glands.
viii. The immune system defends against pathogens, infections, and diseases through a complex network of cells, tissues, and organs.
ix. The urinary system filters waste products and helps regulate water and electrolyte balance, comprising the kidneys, bladder, and associated structures.
x. The reproductive system allows for the production of offspring and includes structures specific to males (testes, penis) and females (ovaries, uterus).

Key Components of the Human Body

The human body is a marvel of biological complexity, composed of various essential elements that contribute to its structure and function. Water, constituting about 60% of the body's weight, is a foundational component crucial for cellular processes, temperature regulation, and nutrient transportation. Proteins, accounting for approximately 16% of body weight, are integral to cell structure, immune function, and as enzymes that facilitate biochemical reactions. Fats, comprising around 16-25% of body weight, play a role in energy storage, insulation, and the construction of cell membranes. Carbohydrates, making up about 1% of body weight, serve as a primary energy source and contribute to cellular structures. Minerals, including calcium, phosphorus, and potassium, are vital for bone health, nerve function, and electrolyte balance. Trace elements like iron and zinc, though present in smaller amounts, are essential for various physiological processes. Vitamins, in minute quantities, function as coenzymes and antioxidants, supporting overall health. Together, these components create a dynamic and intricately balanced system that sustains life and enables the body to perform its myriad functions with precision.

Introduction To Biology

■ Skeletal System

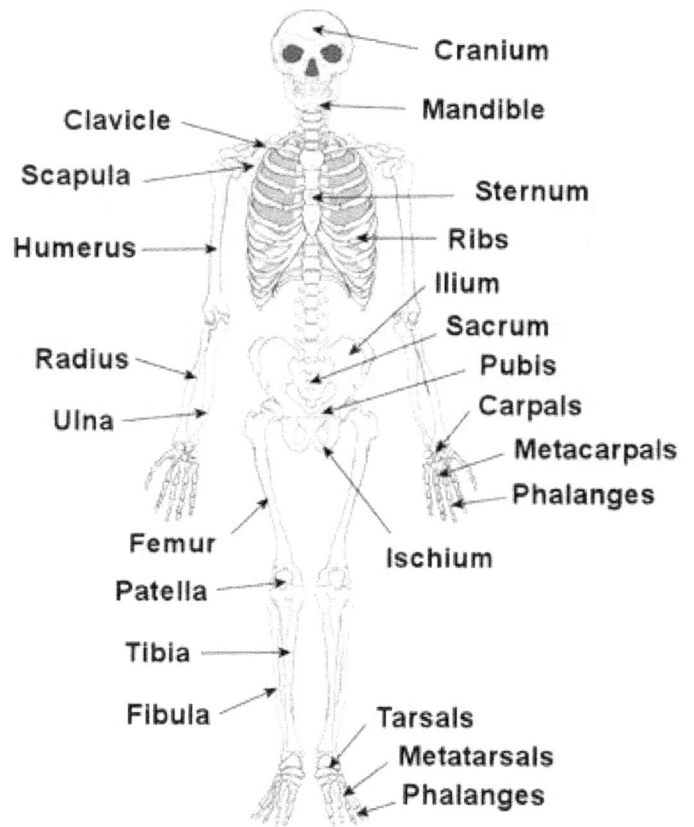

The skeletal system is the framework that provides structure, support, and protection to the human body. Composed of bones, joints, and connective tissues like cartilage and ligaments, this system serves several crucial functions. Firstly, bones offer mechanical support, allowing the body to maintain its shape and withstand external forces. Secondly, the skeletal system plays a vital role in facilitating movement, as muscles attach to bones, creating a system of levers for locomotion.

Bones also protect internal organs; for instance, the ribcage shields the heart and lungs, while the skull guards the brain. Furthermore, bone marrow within certain bones is essential for the production of blood cells. The skeletal system is dynamic and undergoes continuous processes of growth, remodeling, and repair throughout life. Calcium and phosphorus are key minerals stored in bones, contributing to mineral homeostasis in the body. In summary, the skeletal system is indispensable for the overall functionality of the human body, providing a structural framework, enabling movement, and safeguarding vital organs.

Functions Of The Skeletal System

1. **Support:** The skeleton provides structural support to the body, allowing us to stand upright and maintain our body shape.
2. **Protection:** The skeleton protects vital organs such as the brain, spinal cord, heart, and lungs. For example, the skull protects the brain, and the ribcage protects the heart and lungs.
3. **Movement:** The bones, along with the associated muscles, allow for movement. Muscles attach to bones via tendons, and when the muscles contract, they cause the bones to move.
4. **Mineral Storage:** The skeleton acts as a reservoir for minerals, primarily calcium and phosphorus. These minerals can be released into the bloodstream when needed for various bodily functions.
5. **Blood Cell Production:** The bone marrow, found within certain bones, is responsible for producing blood cells, including red blood cells, white blood cells, and platelets.

Bone Structure

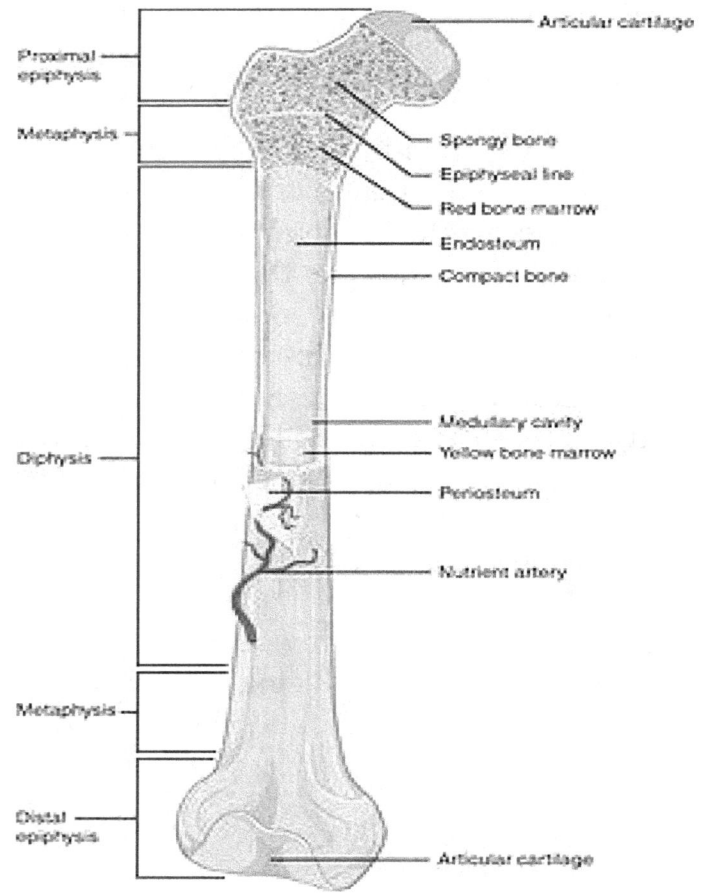

The bone structure is a complex and dynamic framework that provides support, protection, and facilitates bodily movement. Bones are classified into two main types: compact (cortical) bone and spongy (cancellous) bone. The outer layer of bones is composed of compact bone, which is dense and hard, providing strength and resilience. Within this compact bone, there are microscopic channels known as Haversian canals that house blood vessels and nerves, ensuring a continuous supply of nutrients to the bone cells.

Spongy bone, found at the ends of long bones and within flat bones, has a more porous structure, creating a network of trabeculae that adds strength without excessive weight. The central cavity of long bones houses marrow, a critical component for blood cell production and fat storage. Osteocytes, bone cells embedded in the bone matrix, maintain and regulate bone tissue, communicating through tiny channels called canaliculi.

Collagen fibers and mineralized calcium phosphate crystals make up the extracellular matrix, contributing to the bone's resilience and hardness. Bone structure is not static; it undergoes constant remodeling through processes of bone resorption and formation, ensuring adaptability to mechanical stress and maintaining skeletal integrity throughout life. The intricate design of bone structure reflects the remarkable balance between strength, flexibility, and the dynamic nature required for the diverse functions bones serve in the human body.

Joints

Joints, also known as articulations, are pivotal components of the human skeletal system, facilitating movement and providing flexibility to the body. These junctions connect bones and are integral to the structure and function of the musculoskeletal system. There are various types of joints, each with distinct characteristics and ranges of motion.

1. **Fibrous Joints:**

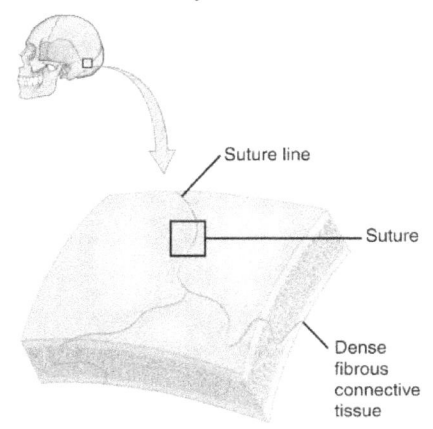

These joints are held together by fibrous connective tissue and allow little to no movement. They provide stability and support to the bones. Examples include sutures in the skull and syndesmosis between the radius and ulna in the forearm.

2. **Cartilaginous Joints:** These joints are connected by cartilage and allow limited movement. There are two types:

i. **Synchondrosis:**

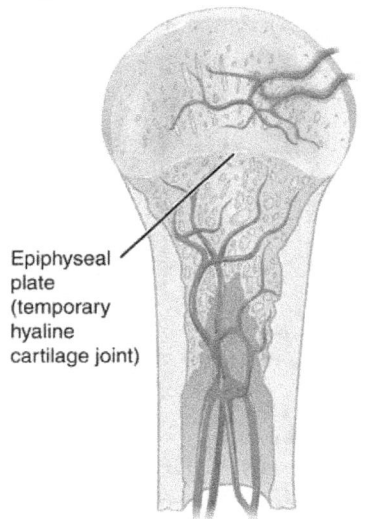

A temporary joint where the connecting tissue is hyaline cartilage. Examples include the growth plates in long bones.

ii. **Symphysis:**

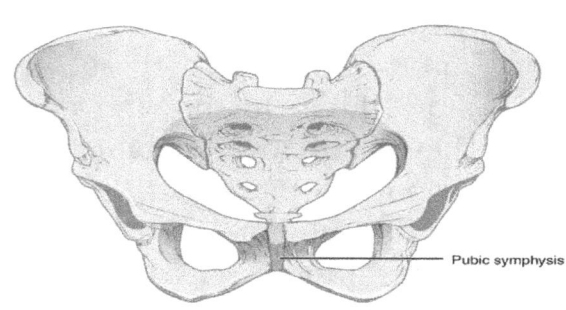

A slightly movable joint where the connecting tissue is fibrocartilage. Examples include the pubic symphysis and intervertebral discs.

3. **Synovial Joints:** These joints are the most common in the body and allow a wide range of movements due to the presence of a synovial cavity filled with synovial fluid. They are classified into several types based on their structure and movement

i. **Ball and Socket Joint:**

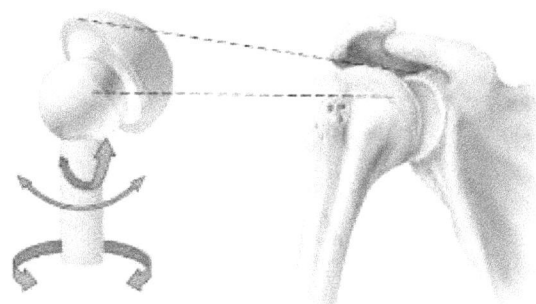

Characterized by a spherical head of one bone fitting into a cup-like socket of another bone. Examples include the shoulder and hip joints.

ii. **Hinge Joint:**

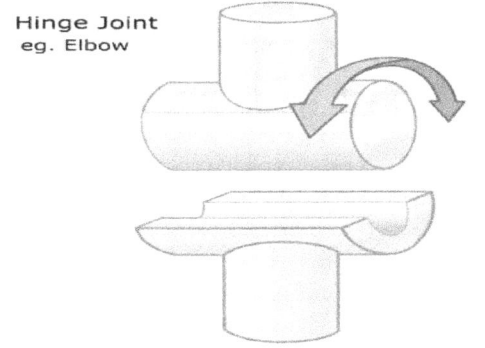

Allows movement in one plane like a door hinge, permitting flexion and extension. Examples include the elbow and knee joints.

iii. **Pivot Joint:**

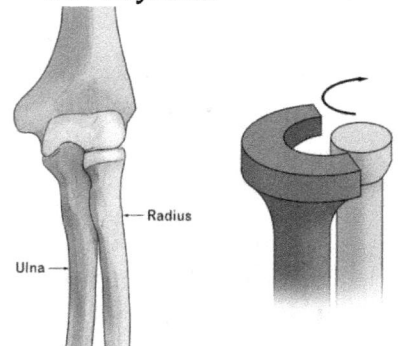

Allows rotation around a central axis. Examples include the joint between the atlas and axis vertebrae in the neck.

iv. **Gliding Joint:**

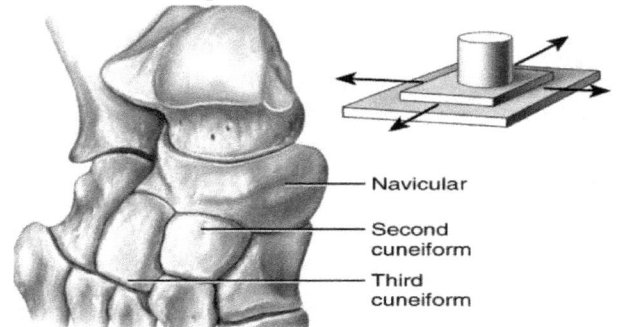

Allows sliding or gliding movements between flat or slightly curved surfaces of bones. Examples include the joints between the carpal and tarsal bones.

v. Saddle Joint:

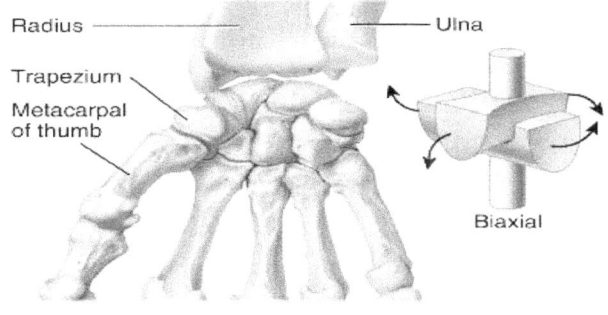

(e) Saddle joint between trapezium of carpus (wrist) and metacarpal of thumb

Characterized by one bone surface concave in one direction and convex in another, allowing movement in two planes. An example is the joint at the base of the thumb.

Cartilage

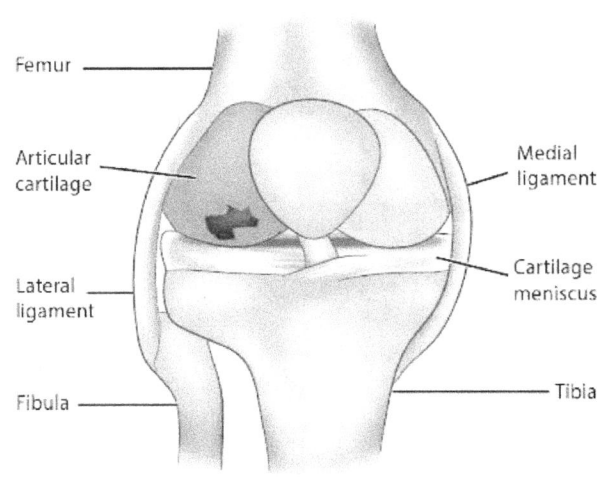

Cartilage is a resilient and flexible connective tissue that plays a crucial role in the human body, providing structural support and facilitating smooth joint movement. Composed of specialized cells called chondrocytes embedded in a matrix of collagen fibers and proteoglycans, cartilage is found in various locations, including the joints, ears, nose, and respiratory passages.

Its unique composition gives cartilage its firm yet pliable nature, making it an excellent shock absorber in joints and providing a smooth surface for bones to glide over during movement. Unlike bone, cartilage lacks blood vessels and nerves, relying on diffusion for nutrient exchange. This avascular nature limits its regenerative capacity, making cartilage repair slower than in other tissues.

Despite this limitation, cartilage is crucial for maintaining joint integrity, supporting the respiratory system, and shaping certain body structures. Its ability to withstand compressive forces and provide structural support makes cartilage an indispensable component in ensuring the functionality and resilience of various anatomical regions within the human body.

Skeletal Disorders

Skeletal disorders encompass a range of conditions that affect the bones, joints, and connective tissues, often impacting the overall structure and function of the skeletal system. Some common skeletal disorders include:

1. **Osteoporosi**

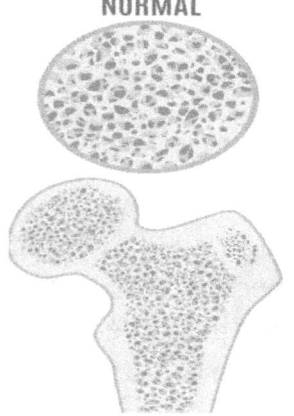 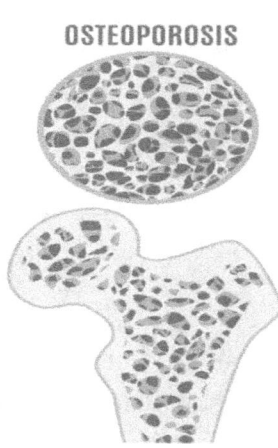

This disorder is characterized by a decrease in bone density and mass, leading to fragile and porous bones. It increases the risk of fractures, especially in postmenopausal women and older adults.

Introduction To Biology

2. ## Arthritis

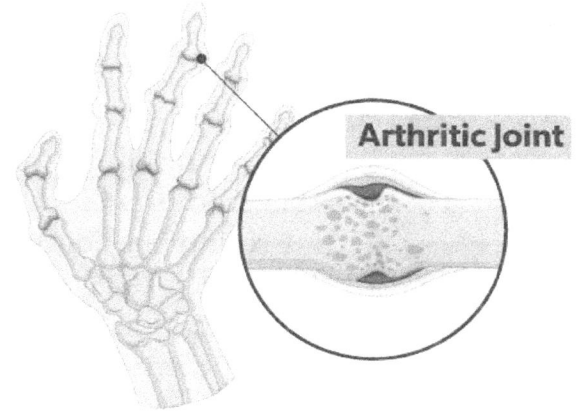

Arthritis refers to inflammation of the joints and is a category of disorders that includes osteoarthritis, rheumatoid arthritis, and juvenile idiopathic arthritis. These conditions result in joint pain, stiffness, and reduced mobility.

3. ## Osteoarthritis

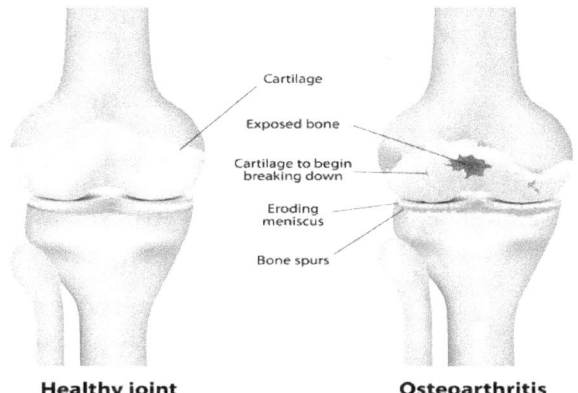

The most prevalent form of arthritis, osteoarthritis involves the gradual degeneration of joint cartilage, leading to pain, swelling, and decreased joint flexibility. It commonly affects weight-bearing joints like the knees and hips.

4. Rheumatoid Arthritis

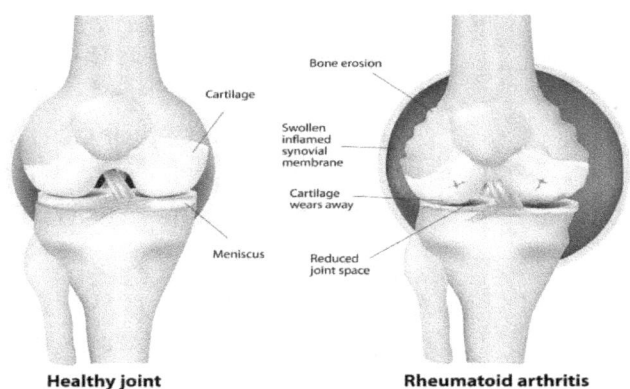

An autoimmune disorder, rheumatoid arthritis causes the immune system to attack the synovium (lining of the membranes that surround the joints), leading to joint inflammation, pain, and potential joint deformities.

5. Osteogenesis Imperfecta (Brittle Bone Disease)

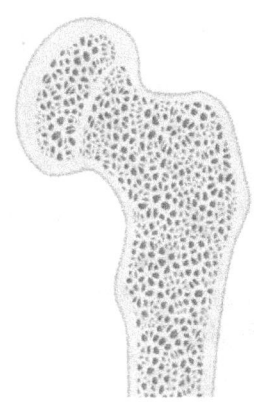
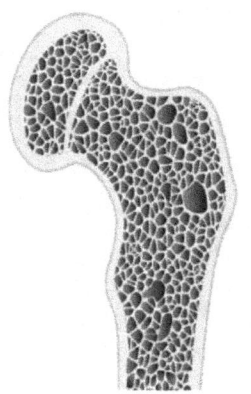

This genetic disorder results in brittle and easily fractured bones due to a deficiency in collagen, a protein essential for bone strength. Individuals with osteogenesis imperfecta often experience frequent fractures and may have a blue tint to the whites of their eyes.

Introduction To Biology

6. Scoliosis

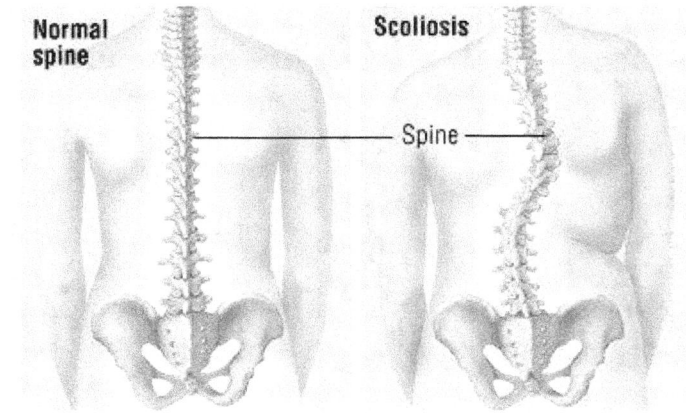

Scoliosis is a condition characterized by an abnormal curvature of the spine. While some cases are mild, severe scoliosis can affect lung and heart function and may require corrective interventions.

7. Kyphosis

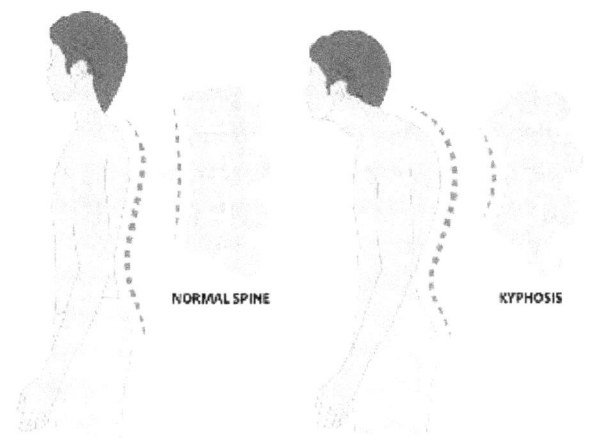

Kyphosis involves an abnormal forward rounding of the upper spine, leading to a hunched or "humpback" appearance. This can be associated with osteoporosis or developmental issues in the spine.

Introduction To Biology

8. Gout

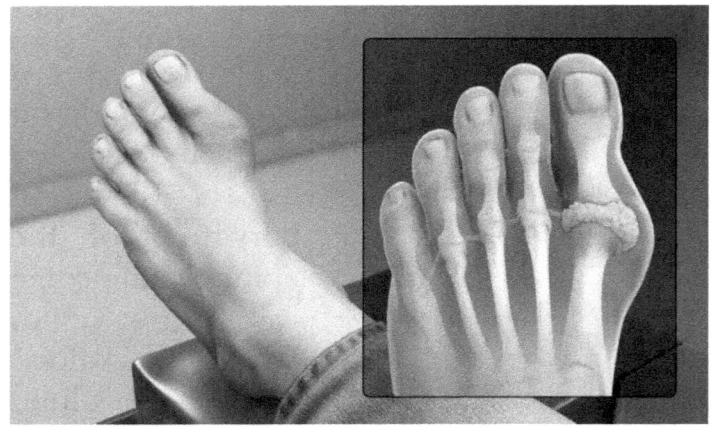

Gout is a form of arthritis caused by the accumulation of urate crystals in the joints, leading to intense pain, swelling, and redness, commonly affecting the big toe.

9. Paget's Disease

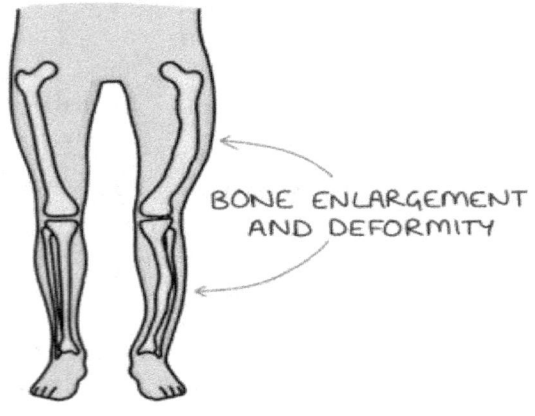

BONE ENLARGEMENT AND DEFORMITY

This disorder results in the abnormal breakdown and formation of bone tissue, leading to enlarged and weakened bones. While often asymptomatic, it can cause pain, deformities, and an increased risk of fractures.

10. Ankylosing Spondylitis

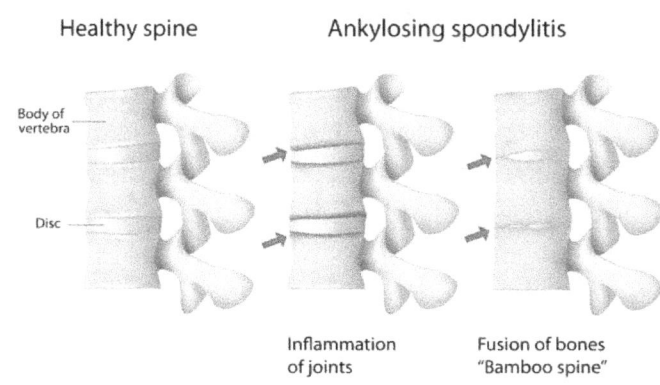

An inflammatory arthritis affecting the spine, ankylosing spondylitis can lead to fusion of the vertebrae, causing stiffness and decreased mobility. It commonly begins in the lower back.

The skeletal system is essential for maintaining the body's form and function. It provides the necessary support for movement, protects vital organs, and plays a crucial role in blood cell production and mineral storage. Understanding the structure and function of the skeletal system is vital for maintaining overall health and well-being.

■ Muscular System

The muscular system is a complex network of tissues and organs responsible for the movement of the body and its various parts. Composed primarily of muscle tissues, the muscular system enables locomotion, maintains posture, generates heat, and facilitates vital functions like breathing and digestion.

Muscle Tissues

The muscular system consists of three types of muscle tissues: skeletal, cardiac, and smooth muscles.

a. **Skeletal muscle:** Also known as striated or voluntary muscle, skeletal muscles are attached to bones and responsible for

voluntary movements. They are composed of long, cylindrical muscle fibers that are multinucleated.
b. **Cardiac muscle:** Found only in the heart, cardiac muscle tissue is involuntary and striated. It allows the heart to contract and pump blood throughout the body. Cardiac muscles are branched and interconnected, forming a coordinated network.
c. **Smooth muscle:** Smooth muscles are found in the walls of internal organs, blood vessels, and the respiratory and digestive systems. They are non-striated, involuntary muscles responsible for automatic movements like peristalsis. Smooth muscles are spindle-shaped and have a single nucleus.

Types of Muscle

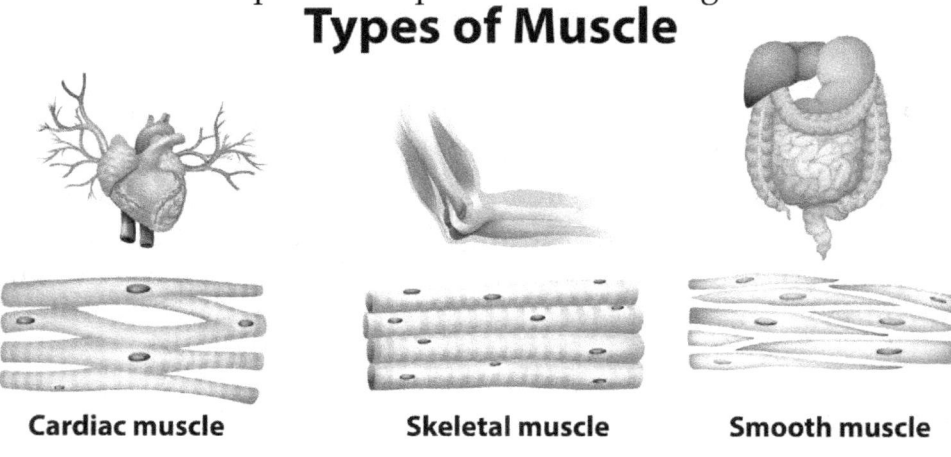

Cardiac muscle Skeletal muscle Smooth muscle

Functions Of Muscles

1. **Movement:** Muscles allow for voluntary and involuntary movements of the body and its various parts. Skeletal muscles work together with bones and joints to produce coordinated movements like walking, running, and lifting objects.
2. **Posture and Stability:** Muscles play a crucial role in maintaining posture and stability. They provide support to the skeletal system, keeping the body upright and balanced against gravity.

3. **Heat Production:** Muscle contractions generate heat, which helps regulate body temperature. This is particularly important during physical activity or in cold environments.
4. **Protection of Internal Organs:** Certain muscles, like the abdominal and pelvic muscles, help protect vital organs by surrounding and supporting them.
5. **Control of Openings:** Muscles control the openings of various body passages, such as the mouth, anus, and urinary sphincters, allowing for voluntary control over functions like eating, elimination, and urination.
6. **Facilitation of Organ Functions:** Smooth muscles within the walls of organs, such as the intestines and blood vessels, facilitate the movement of substances, such as food and blood, through these structures.

Muscle Structure

Muscles are composed of individual muscle fibers bundled together by connective tissues. Each muscle fiber contains smaller units called myofibrils, which are composed of even smaller units called sarcomeres. Sarcomeres are the functional units responsible for muscle contraction. Within the sarcomeres, two types of protein filaments—actin and myosin—interact to produce muscle contractions. Muscles are connected to bones via tendons, which are strong, fibrous connective tissues that transmit the force generated by muscle contractions to produce movement.

Muscle Contraction

Muscle contraction occurs when actin and myosin filaments within sarcomeres slide past each other, causing the sarcomere to shorten. Contraction is initiated by nerve impulses sent from the brain and spinal cord to the muscles, leading to the release of calcium ions and the formation of cross-bridges between actin and myosin filaments. The interaction between actin and myosin filaments is powered by the hydrolysis of ATP (adenosine triphosphate), which provides the necessary energy for muscle contraction.

Muscle Control and Coordination

The muscular system is regulated by the nervous system, which sends signals from the brain and spinal cord to initiate and control muscle contractions. Motor neurons transmit these signals to the muscle fibers, leading to their activation and contraction. The coordinated action of multiple muscles is necessary for smooth and precise movements, which is achieved through the integration of motor signals and feedback from sensory receptors.

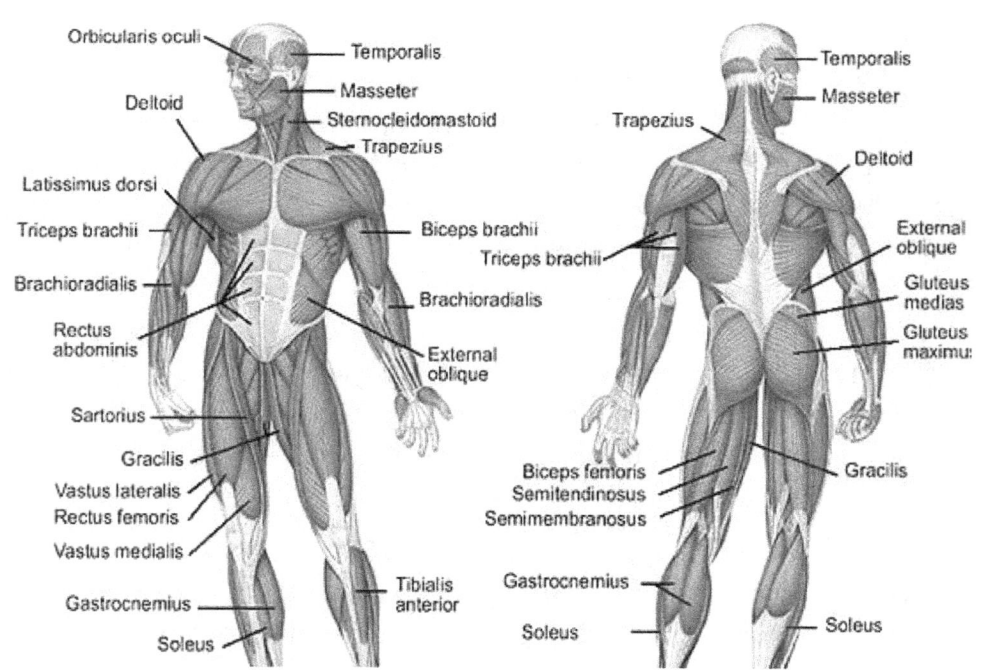

Nervous System

The nervous system is a complex network of cells and tissues that enables communication and coordination throughout the body. It plays a crucial role in regulating and controlling various bodily functions, including sensory perception, motor control, cognition, and behavior. The nervous system is protected by various structures, including the skull, vertebral column, meninges (protective membranes around the brain and spinal cord), and cerebrospinal fluid (which cushions and nourishes the CNS).

Structure of the Nervous System

1. **Central Nervous System (CNS):** The CNS consists of the brain and spinal cord. It serves as the command center of the nervous system, processing information and coordinating responses.
2. **Peripheral Nervous System (PNS):** The PNS comprises nerves and ganglia outside the CNS. It connects the CNS to the rest of the body and transmits information between the CNS and peripheral organs.

Neurons

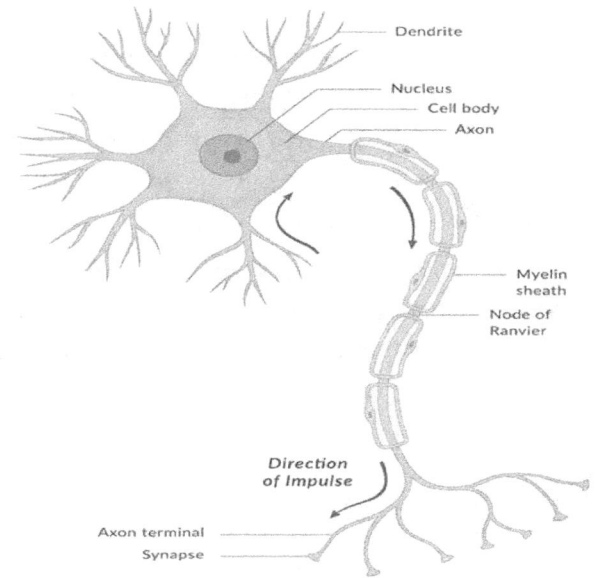

Neurons are specialized cells that transmit electrical signals called nerve impulses or action potentials. They consist of a cell body, dendrites (extensions that receive signals), and an axon (a long projection that transmits signals to other neurons or target cells). Neurons communicate with each other at specialized junctions called synapses, where signals are transmitted chemically across the gaps between neurons.

Types of Neurons

1. **Sensory Neurons:** These neurons transmit sensory information from sensory organs (such as the eyes, ears, and skin) to the CNS.
2. **Motor Neurons:** Motor neurons carry signals from the CNS to muscles and glands, controlling movement and initiating responses.
3. **Interneurons:** Interneurons form connections between sensory and motor neurons in the CNS, allowing for complex processing and integration of information.

Divisions of the Nervous System

a. **Somatic Nervous System:** The somatic nervous system controls voluntary movements and receives sensory information from external stimuli.
b. **Autonomic Nervous System (ANS):** The ANS regulates involuntary functions of the body, including heart rate, digestion, and respiration. It is further divided into the sympathetic and parasympathetic divisions, which have opposing effects on bodily functions.

Central Nervous System (CNS)

The brain is the command center of the nervous system. It is divided into regions responsible for different functions, including sensory perception, motor control, memory, emotions, and cognition. The spinal cord connects the brain to the PNS and controls reflexes and the transmission of sensory and motor signals.

Peripheral Nervous System (PNS):

The PNS includes cranial nerves and spinal nerves that transmit information between the CNS and the rest of the body. Cranial nerves control functions of the head, neck, and face, such as vision, hearing, and facial movements. Spinal nerves connect the spinal cord to specific regions of the body, controlling movements and transmitting sensory information.

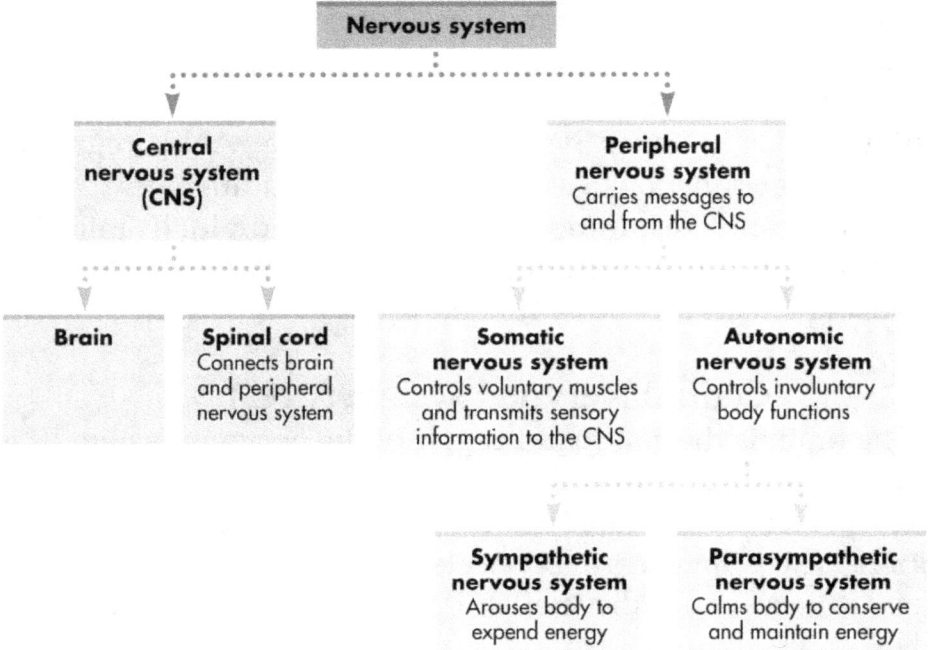

Neurotransmitters: Neurotransmitters are chemical messengers that transmit signals across synapses. Examples of neurotransmitters include serotonin, dopamine, acetylcholine, and norepinephrine, which play essential roles in mood regulation, motor control, and cognitive functions.

Disorders And Diseases

Various disorders and diseases can affect the nervous system, including neurodegenerative diseases (e.g., Alzheimer's, Parkinson's), stroke, epilepsy, multiple sclerosis, and psychiatric disorders.

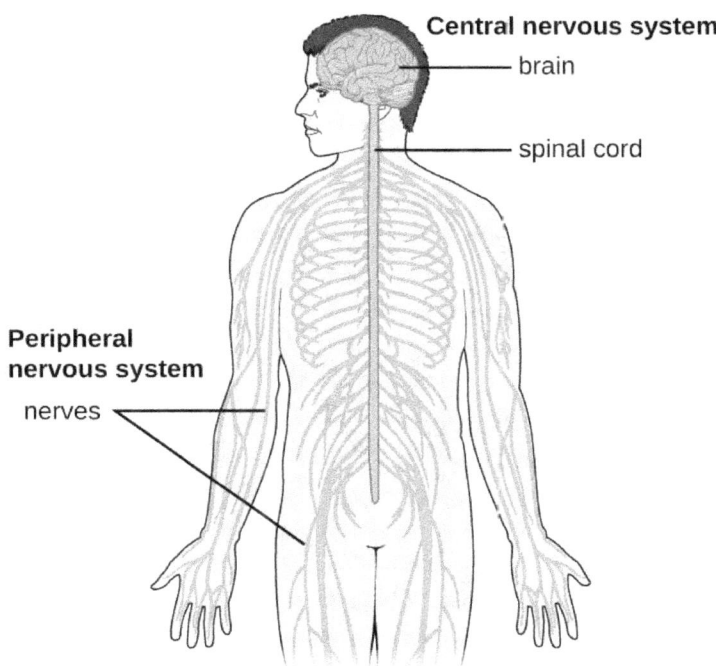

- # Circulatory System

The circulatory system, also known as the cardiovascular system, is a complex network of blood vessels, the heart, and blood that collaboratively ensures the delivery of oxygen, nutrients, and hormones to cells throughout the body while facilitating the removal of waste products. This intricate system plays a pivotal role in maintaining homeostasis, supporting the immune response, and regulating temperature.

Heart

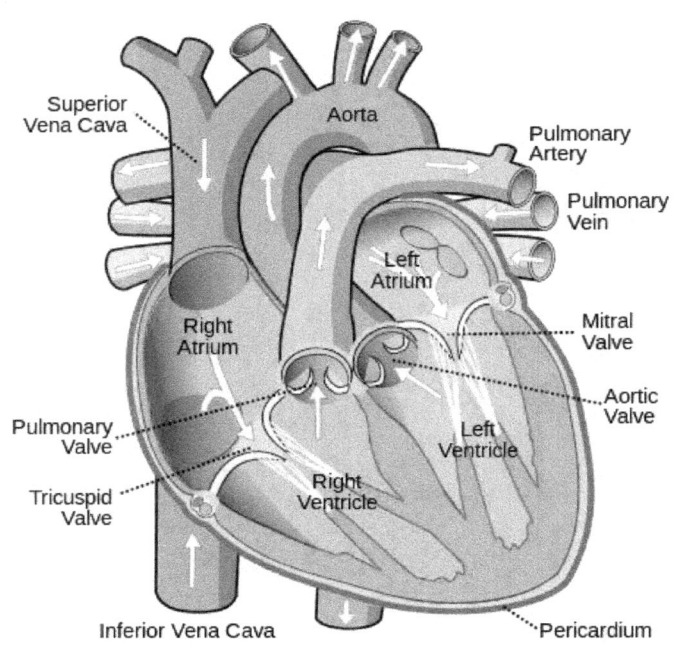

The heart, a vital organ at the core of the circulatory system, is a muscular pump that plays a central role in maintaining blood circulation throughout the body. Located slightly left of the center of the chest, the heart consists of four chambers: two atria (upper chambers) and two ventricles (lower chambers). The right atrium receives deoxygenated blood from the body through the superior and inferior vena cava, while the left atrium receives oxygenated blood from the lungs via the pulmonary veins. The heart's rhythmic contractions, facilitated by the coordinated action of electrical impulses, propel blood through a series of intricate pathways.

The right ventricle pumps deoxygenated blood into the pulmonary arteries, initiating the pulmonary circulation that carries blood to the lungs for oxygenation. Oxygenated blood then returns to the left atrium through the pulmonary veins. The left ventricle, being the more muscular chamber, forcefully contracts to propel oxygenated blood into the aorta, initiating the systemic circulation that delivers oxygen and nutrients to tissues throughout the body. Valves within the heart, such as the atrioventricular (AV) valves and semilunar valves, ensure a unidirectional flow of blood, preventing backflow and maintaining the efficiency of the circulation.

The heart's ability to adjust its rate and force of contraction in response to the body's demands is regulated by the autonomic nervous system and hormonal signals. The sinoatrial (SA) node, often referred to as the heart's natural pacemaker, initiates the electrical impulses that coordinate the heart's contractions. The cardiac cycle, comprising systole (contraction) and diastole (relaxation) phases, ensures a continuous and coordinated flow of blood. This dynamic process is essential for delivering oxygen and nutrients to cells and removing waste products.

Blood Vessels

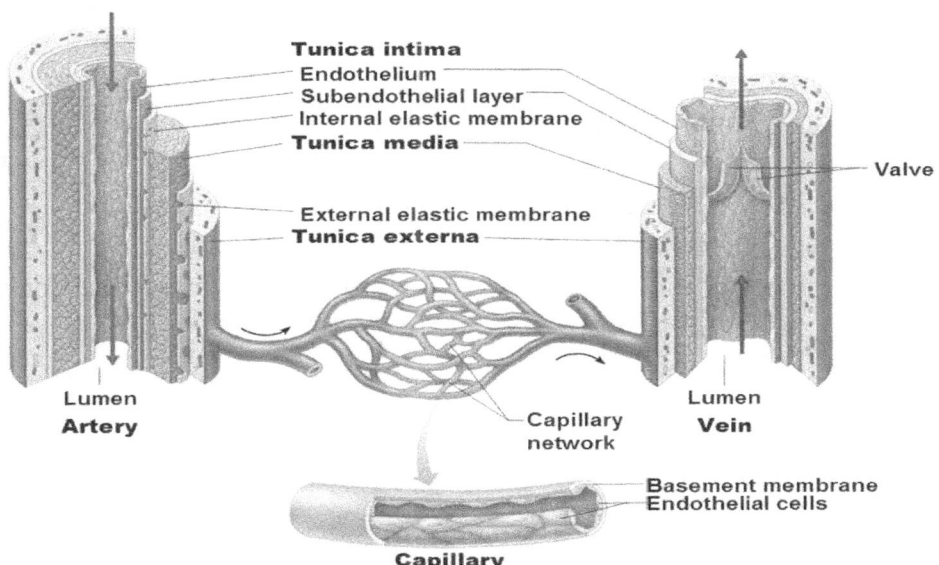

Blood vessels are integral components of the circulatory system, forming an extensive network that transports blood throughout the body, facilitating the delivery of essential nutrients, oxygen, and hormones to tissues while removing waste products. The three main types of blood vessels are arteries, veins, and capillaries, each with distinctive structures and functions.

- **Arteries:** Arteries carry oxygenated blood away from the heart to various tissues. The largest artery, the aorta, branches into smaller arteries that further divide into arterioles. Arteries are characterized by thick, elastic walls that withstand the force exerted by the heart's rhythmic contractions. The elasticity allows arteries to expand and recoil, helping maintain consistent blood flow and pressure throughout the circulatory system.

- **Veins:** Veins transport deoxygenated blood back to the heart. The smallest veins, venules, converge into larger veins that eventually lead to the superior and inferior vena cava, returning blood to the right atrium of the heart. Unlike arteries, veins have thinner walls and less elastic tissue. To counteract the force of gravity and prevent backflow, veins contain one-way valves that facilitate the unidirectional flow of blood toward the heart.

- **Capillaries:** Capillaries are the smallest and most numerous blood vessels, connecting arteries and veins. Their thin walls allow for the exchange of gases, nutrients, and waste products between the blood and surrounding tissues. Capillaries are crucial for the nourishment and oxygenation of cells. The network of capillaries is so extensive that no cell in the body is far from these microscopic vessels.

Microcirculation: The intricate system of arteries, veins, and capillaries collectively forms microcirculation. This complex network ensures efficient nutrient exchange and waste removal at the cellular level. Blood flows from arteries to arterioles, then through capillaries, where exchange occurs, and finally, blood returns to the heart through venules and veins.

Blood

Blood is a specialized connective tissue that flows through the circulatory system. It is a fundamental component of the human circulatory system, playing a crucial role in transporting oxygen, nutrients, hormones, and waste products throughout the body. Comprising approximately 7-8% of total body weight, blood is composed of several elements, each with distinct functions. Blood cells are produced in the bone marrow through a process called hematopoiesis. Red blood cells have a limited lifespan and are continually replaced, while white blood cells and platelets are also generated as needed. Old or damaged blood cells are removed by the spleen and liver.

➢ **Plasma:** The liquid component of blood, known as plasma, makes up about 55% of its volume. Plasma is a yellowish fluid consisting mainly of water, electrolytes, proteins, hormones, and waste products. It serves as a transport medium, carrying cells, nutrients, and other substances to various tissues.
➢ **Red Blood Cells (Erythrocytes):**

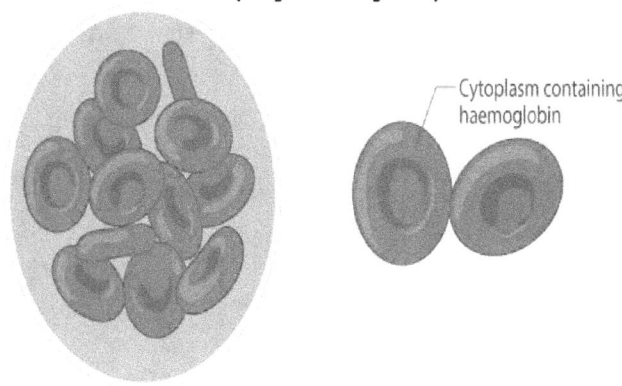

Making up the majority of blood cells, red blood cells make up about 38-45% of the total blood volume. The normal range for red blood cell count is around 4.5 to 6 million cells per microliter for adult males and 4 to 5.5

million cells per microliter for adult females. red blood cells are specialized for oxygen transport. They contain the protein hemoglobin, which binds to oxygen in the lungs and releases it in tissues, ensuring efficient oxygenation of body cells. The lack of a nucleus allows red blood cells to have a unique biconcave shape, enhancing their flexibility and surface area.

> **White Blood Cells (Leukocytes):**

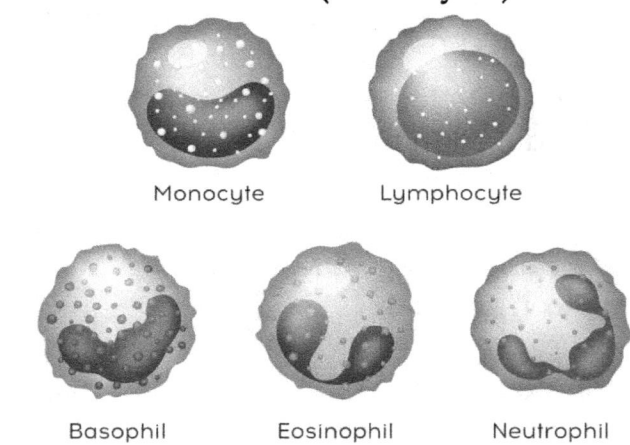

White blood cells account for a smaller percentage, roughly 1% of the total blood volume. The total white blood cell count is typically between 4,000 and 11,000 cells per microliter. White blood cells play a vital role in the immune system, defending the body against infections and foreign substances. They are divided into several types, including neutrophils, lymphocytes, monocytes, eosinophils, and basophils, each serving specific functions in immune response and defense mechanisms.

Introduction To Biology

➤ **Platelets (Thrombocytes):**

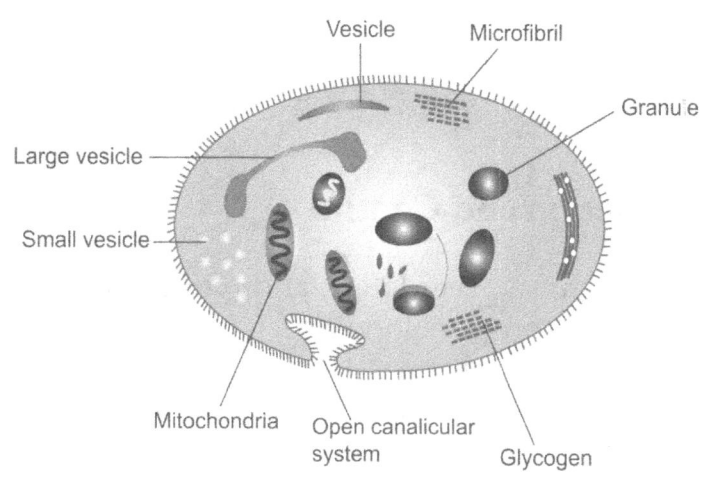

Platelets constitute a very small portion of the blood, making up about 0.1% of the total volume. The normal platelet count ranges from 150,000 to 450,000 platelets per microliter. Small cell fragments, platelets contribute to blood clotting and coagulation. When a blood vessel is injured, platelets adhere to the site, release clotting factors, and form a plug to prevent excessive bleeding.

Circulation

Circulation is a fundamental physiological process that orchestrates the movement of blood, nutrients, gases, and waste products throughout the circulatory system. The heart, a muscular organ, propels blood through a dual circulation system, involving both pulmonary and systemic pathways. Arteries, arterioles, capillaries, venules, and veins form an intricate network facilitating the seamless flow of blood. As the heart contracts, oxygenated blood is pumped into the aorta, circulating through arteries, capillaries, and veins.

The blood delivers oxygen and nutrients to tissues via capillaries, where exchange occurs, and waste products enter the bloodstream. Venules and veins then return deoxygenated blood to the heart, completing the cyclical process. The autonomic nervous system and hormones regulate circulation, adjusting

Introduction To Biology

heart rate, vessel diameter, and blood flow to meet the body's changing needs. In essence, circulation is vital for sustaining life, delivering essential resources to cells, regulating body functions, and contributing to the overall health of the human body.

Regulation and Control

The circulatory system is regulated and controlled by various mechanisms to ensure proper blood flow and distribution. The heart rate and strength of contractions are controlled by electrical signals generated by the sinoatrial (SA) node and the atrioventricular (AV) node, collectively known as the cardiac conduction system. Blood pressure is regulated by the constriction and relaxation of the blood vessels and is influenced by hormones and the autonomic nervous system. Oxygen and carbon dioxide levels in the blood are monitored by chemoreceptors, which help regulate breathing and blood flow.

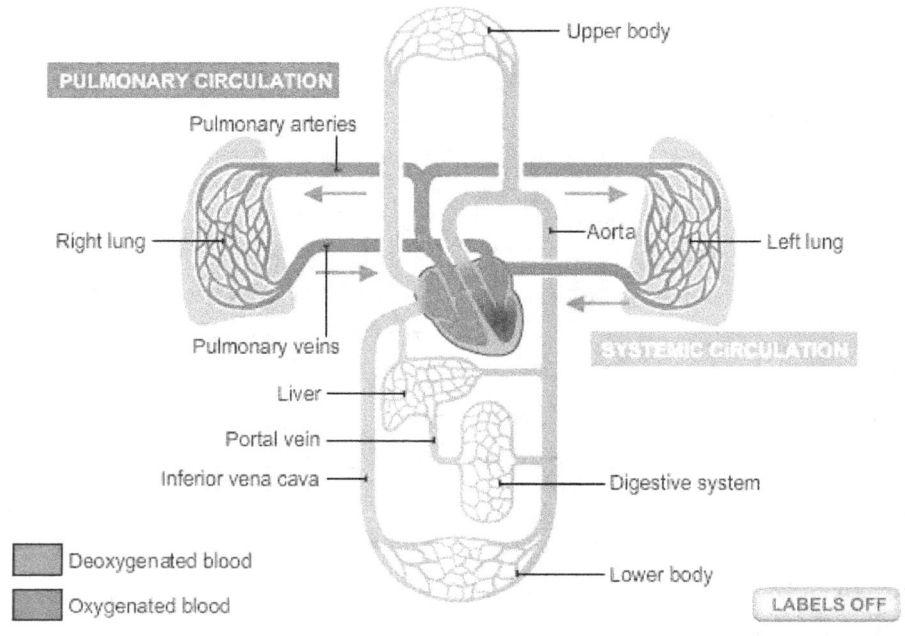

Respiratory System

The respiratory system is a complex network of organs and tissues responsible for the exchange of gases, primarily oxygen and carbon dioxide, between the body and the environment. It plays a crucial role in supporting cellular respiration, providing the oxygen necessary for energy production and removing carbon dioxide as a waste product. Comprising both upper and lower respiratory tracts, this intricate system ensures the continuous flow of air into and out of the body.

a. **Upper Respiratory Tract:** The upper respiratory tract includes the nose, nasal cavity, pharynx, and larynx. The nose, with its hair-lined passages, filters and humidifies incoming air, while the pharynx and larynx guide air to the lower respiratory tract and help produce sound during speech.
- **Nasal Cavity:** The respiratory process begins in the nasal cavity, where air enters the body through the nostrils. The nasal cavity is lined with a mucous membrane that helps filter, warm, and moisten the incoming air. Fine hairs called cilia and mucus trap dust, pollen, and other airborne particles, preventing them from reaching the lungs.
- **Pharynx and Larynx:** From the nasal cavity, air passes through the pharynx, a muscular tube shared with the digestive system. The larynx, or voice box, is located at the top of the trachea and contains vocal cords responsible for sound production. The epiglottis, a flap of tissue, covers the entrance to the larynx during swallowing to prevent food from entering the respiratory tract.

b. **Lower Respiratory Tract:** The lower respiratory tract involves the trachea, bronchi, bronchioles, and lungs. The trachea, supported by cartilage rings, carries air from the larynx to the bronchi, which further divide into bronchioles. These bronchioles lead to tiny air sacs called alveoli, where the crucial exchange of gases takes place.

✓ **Trachea and Bronchial Tree:** The trachea, or windpipe, is a rigid tube composed of cartilage rings that extends from the larynx to the chest. Within the chest, the trachea divides into two main bronchi, one entering each lung. The bronchi further divide into smaller bronchioles, forming the bronchial tree. The walls of the bronchial tree contain smooth muscle that can contract or relax, controlling the flow of air.

✓ **Lungs:** The lungs are the main respiratory organs and are located in the chest cavity on either side of the heart. Each lung is surrounded by a double-layered membrane called the pleura, which helps reduce friction during breathing. The lungs are composed of millions of tiny air sacs called alveoli, where gas exchange occurs.

✓ **Alveoli and Gas Exchange:** The alveoli are surrounded by an extensive network of capillaries, allowing for efficient gas exchange. Oxygen from inhaled air diffuses across the thin alveolar walls into the bloodstream, binding to hemoglobin in red blood cells. Carbon dioxide, a waste product of cellular respiration, diffuses from the bloodstream into the alveoli to be exhaled.

Diaphragm And Breathing

The diaphragm is a crucial muscle involved in the process of breathing, playing a central role in the expansion and contraction of the chest cavity. This dome-shaped muscle separates the thoracic cavity, housing the heart and lungs, from the abdominal cavity. As a primary respiratory muscle, the diaphragm is essential for the inhalation and exhalation of air.

During inhalation, the diaphragm contracts and moves downward, creating a larger space in the thoracic cavity. Simultaneously, the intercostal muscles between the ribs contract, lifting and expanding the ribcage. This expansion lowers the air pressure in the lungs, causing air to rush in and fill the vacuum created. This entire process is known as negative pressure breathing, where the drop in pressure draws air into the lungs.

Conversely, during exhalation, the diaphragm relaxes and moves upward, while the intercostal muscles relax, allowing the ribcage to decrease in size. This compression increases the air pressure in the lungs, causing air to be expelled. Exhalation is generally a passive process, relying on the elasticity of the lung tissue and the relaxation of muscles.

The diaphragm's role in breathing is under involuntary control, primarily regulated by the autonomic nervous system. However, it can also be influenced voluntarily through activities such as deep breathing or diaphragmatic breathing exercises. These exercises aim to engage the diaphragm more effectively, promoting better lung capacity and respiratory efficiency. Various factors, including physical activity, respiratory conditions, and emotional states, can impact the function of the diaphragm and breathing. Conditions like asthma, chronic obstructive pulmonary disease (COPD), or anxiety disorders can affect the diaphragm's ability to contract and relax efficiently, leading to breathing difficulties.

Respiratory Control

Respiratory control is a complex physiological process that regulates the rate and depth of breathing to ensure the efficient exchange of gases, primarily oxygen and carbon dioxide, in the body. This intricate system involves various sensors, neural pathways, and feedback mechanisms that work together to maintain the balance of oxygen and carbon dioxide levels in the bloodstream.

1. **Central Respiratory Control:** At the core of respiratory control is the central respiratory center located in the brainstem, particularly in the medulla oblongata and the pons. This center receives input from peripheral chemoreceptors and other sensory receptors, integrating this information to regulate respiratory patterns. The medullary respiratory center, consisting of the dorsal respiratory group and the ventral respiratory group, plays a crucial role in coordinating the inspiratory and expiratory phases of breathing.
2. **Peripheral Chemoreceptors:** Peripheral chemoreceptors, primarily located in the carotid bodies and the aortic bodies, sense changes in the partial pressure of oxygen (PaO2), carbon dioxide (PaCO2), and pH in the blood. When oxygen levels decrease or carbon dioxide levels rise, these receptors send signals to the central respiratory center to adjust breathing rates and depths accordingly.
3. **Central Chemoreceptors:** Central chemoreceptors, found near the ventral surface of the medulla, are sensitive to changes in the pH of the cerebrospinal fluid. While they do not directly respond to oxygen levels, an increase in carbon dioxide levels in the blood leads to the production of carbonic acid, lowering the pH of the cerebrospinal fluid. This drop in pH stimulates the central chemoreceptors, signaling an increase in respiratory rate to expel excess carbon dioxide and restore pH balance.
4. **Mechanoreceptors and Lung Stretch Receptors:** Mechanoreceptors in the lungs and chest wall play a role in respiratory control by sensing changes in lung volume and the stretch of lung tissues. These receptors provide feedback to the central respiratory center, influencing the timing and duration

of each respiratory cycle. The Hering-Breuer reflex, activated by stretch receptors in the lungs, helps prevent overinflation during inspiration.
5. **Integration and Feedback Mechanisms:** Respiratory control involves continuous integration of information from multiple sources, including peripheral and central chemoreceptors, mechanoreceptors, and other sensory inputs. The central respiratory center adjusts breathing patterns based on these inputs to maintain homeostasis in oxygen and carbon dioxide levels.

Various factors can influence respiratory control, including metabolic demands, physical activity, emotional states, and external stimuli. For example, increased metabolic activity or heightened emotional states may lead to an increase in respiratory rate to meet the body's oxygen demands.

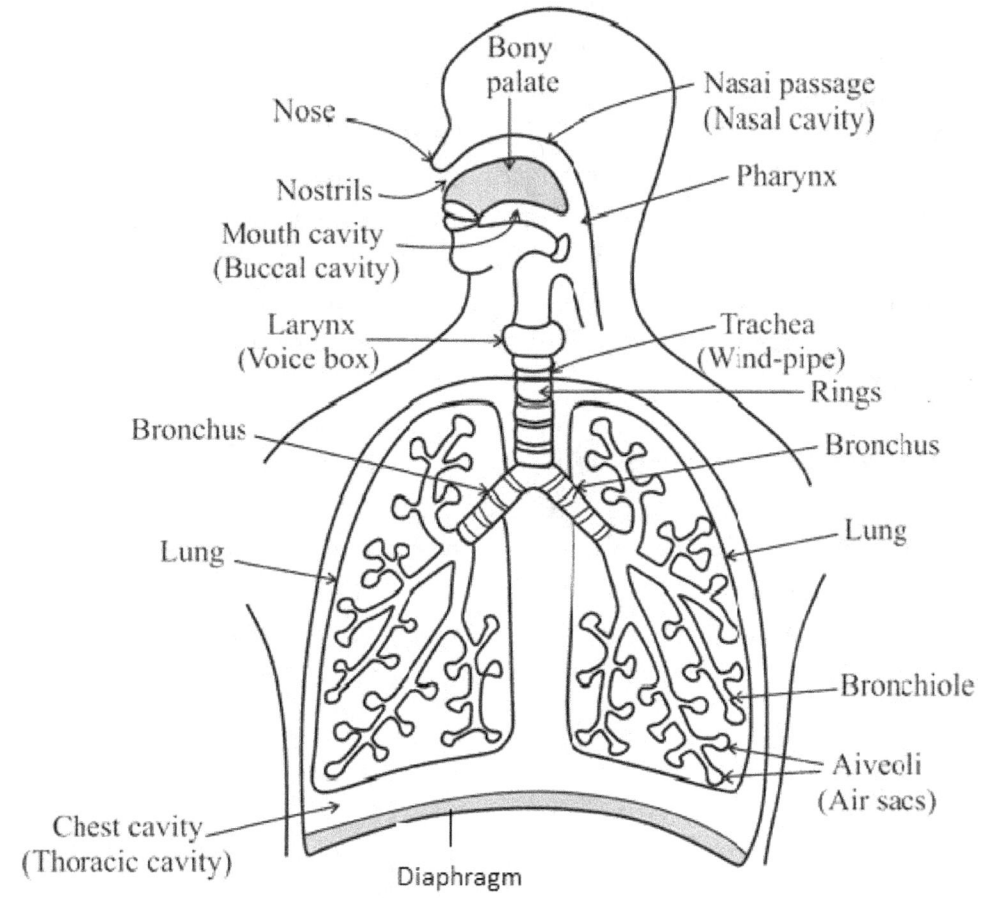

■ Digestive System

The digestive system is a complex series of organs and processes that work together to break down food, extract nutrients, and eliminate waste from the body. It plays a crucial role in providing the body with energy and essential nutrients for growth, repair, and maintenance.

- **Mouth:** The process of digestion begins in the mouth, where food is ingested and mechanically broken down by chewing and mixing with saliva. Saliva, produced by salivary glands, contains enzymes (such as amylase) that initiate the breakdown of carbohydrates.
- **Pharynx and Esophagus:** After chewing, the food is formed into a bolus and swallowed. The bolus travels through the pharynx (throat) and enters the esophagus, a muscular tube that carries the food to the stomach through rhythmic contractions called peristalsis.
- **Stomach:** The stomach is a muscular organ that continues the mechanical and chemical digestion of food. It secretes gastric juices, including hydrochloric acid and enzymes (such as pepsin), which break down proteins. The stomach's muscular contractions help mix and churn the food, forming a semi-liquid substance called chyme.
- **Small Intestine:** The small intestine is the primary site for digestion and absorption of nutrients. It receives chyme from the stomach and is approximately 6 meters long. The inner lining of the small intestine has numerous folds and finger-like projections called villi, which increase the surface area for nutrient absorption.
- Enzymes produced by the pancreas and intestinal wall break down carbohydrates, proteins, and fats into smaller molecules. Nutrients, such as sugars, amino acids, and fatty acids, are absorbed into the bloodstream through the villi.
- **Liver, Gallbladder, and Pancreas:** The liver produces bile, which is stored and concentrated in the gallbladder. Bile aids in the digestion and absorption of fats. The pancreas produces digestive enzymes (amylase, lipase, proteases) and releases them into the small intestine to further break down nutrients.
- **Large Intestine:** The large intestine, also known as the colon, absorbs water and electrolytes from the remaining indigestible

food matter. Beneficial bacteria in the colon help break down certain substances and produce vitamins. The colon also plays a role in the formation and elimination of feces through the rectum and anus.

➤ **Rectum and Anus:** The rectum stores feces until it is eliminated from the body through the anus during defecation.

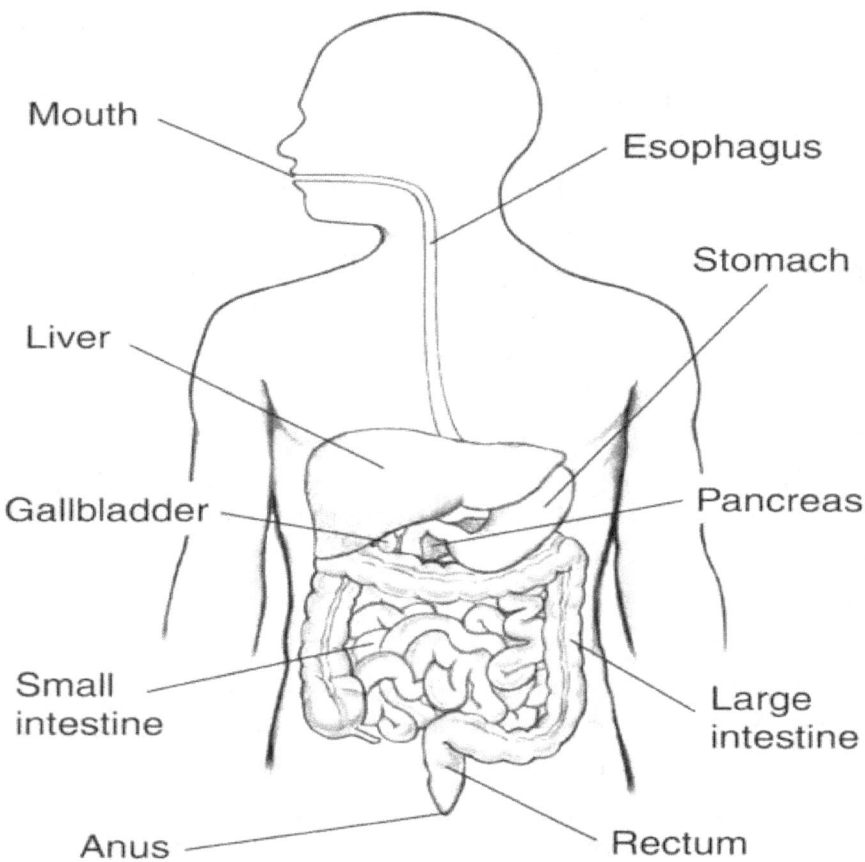

Excretory System

The excretory system is responsible for the elimination of waste products from the body and the maintenance of fluid and electrolyte balance. It includes several organs and structures that work together to filter the blood, remove waste substances, regulate the concentration of various ions, and control the volume and composition of urine.

The excretory system not only eliminates waste products but also helps regulate the balance of water, electrolytes, and pH levels in the body. It works in coordination with other systems, such as the circulatory system, to maintain homeostasis. Proper functioning of the excretory system is crucial for overall health and the elimination of harmful substances from the body.

- **Kidneys:** The kidneys are the primary organs of the excretory system. Humans typically have two kidneys, located on either side of the spine, in the upper abdominal cavity. The kidneys filter the blood and produce urine. They contain millions of tiny filtering units called nephrons, which are responsible for the filtration, reabsorption, and secretion processes.
- ✓ **Filtration:** Blood enters the kidneys through renal arteries, and the nephrons filter out waste products, excess ions, and water from the blood, forming a fluid called filtrate.
- ✓ **Reabsorption:** Useful substances such as water, glucose, ions, and amino acids are reabsorbed from the filtrate and returned to the bloodstream.
- ✓ **Secretion:** Waste products, drugs, and excess ions that were not filtered out during the initial filtration process are actively secreted into the filtrate.
- ✓ The remaining fluid, called urine, is then collected in the renal pelvis and transported to the bladder for storage and subsequent elimination.

Urinary Bladder: The urinary bladder is a hollow, muscular organ located in the pelvis. It serves as a temporary reservoir for urine before it is eliminated from the body. The bladder can expand and contract to accommodate varying volumes of urine. When the bladder reaches a certain level of distention, nerve signals trigger the urge to urinate, and the urine is expelled from the body through the urethra.

Ureters: The ureters are thin, muscular tubes that connect the kidneys to the bladder. They transport urine from the kidneys to the bladder through peristaltic contractions, which are rhythmic muscular movements that help propel urine forward.

Urethra: The urethra is a tube that carries urine from the bladder to the exterior of the body. In males, the urethra serves a dual function by also conducting semen during ejaculation.

Other excretory organs and structures:
- ✓ **Skin:** The skin plays a role in excretion by eliminating small amounts of waste products, such as water, salts, and urea, through sweat glands.
- ✓ **Lungs:** The lungs excrete carbon dioxide and water vapor during respiration.

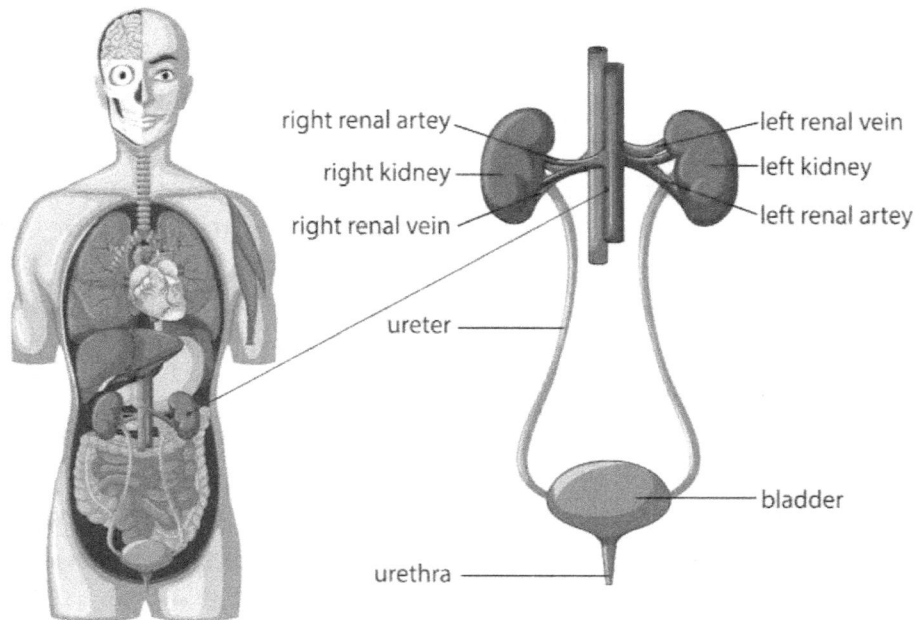

■ Reproductive System

The reproductive system is an essential system in organisms responsible for the production of offspring and the continuation of species. It is a complex system that varies between different organisms but shares the common goal of ensuring successful reproduction. Let's explore the reproductive system in more detail:

Male Reproductive System

The male reproductive system is a complex and highly specialized network of organs and structures responsible for the production, transport, and delivery of sperm, as well as the secretion of hormones vital for male reproductive function. This system plays a central role in sexual reproduction.

i. **Testes:** The primary male reproductive organs are the testes, which are located in the scrotum. The testes serve two essential functions: the production of sperm through a process called spermatogenesis and the secretion of testosterone, the primary male sex hormone.

ii. **Epididymis:** The testes are connected to the epididymis, a coiled tube where sperm mature and gain the ability to swim and fertilize an egg.

iii. **Vas Deferens:** The vas deferens is a muscular tube that transports mature sperm from the epididymis to the ejaculatory duct during ejaculation.

iv. **Seminal Vesicles:** These glands secrete a fluid rich in fructose and other substances that provide energy for the sperm. This fluid, when combined with sperm, forms semen.

v. **Prostate Gland:** The prostate gland produces a milky fluid that contributes to semen's composition. It also surrounds the urethra, playing a role in controlling the flow of urine and semen.

vi. **Bulbourethral Glands:** Also known as Cowper's glands, these small glands secrete a clear fluid that lubricates the urethra and neutralizes any acidic urine residues, creating a more favorable environment for sperm.

vii. **Urethra:** The urethra is a duct that runs through the penis, serving as a passage for both urine and semen. During ejaculation, sperm and seminal fluids exit the body through the urethra.

viii. **Penis:** The penis is an external organ with erectile tissue that becomes engorged with blood during sexual arousal, leading to an erection. The penis facilitates the delivery of sperm into the female reproductive tract during sexual intercourse.

Hormonal Regulation: The hypothalamus, pituitary gland, and testes collaborate in a complex endocrine system to regulate male reproductive functions. The hypothalamus secretes gonadotropin-releasing hormone (GnRH), which stimulates the pituitary gland to release luteinizing hormone (LH) and follicle-stimulating hormone (FSH). LH and FSH then stimulate the testes to produce testosterone and initiate spematogenesis

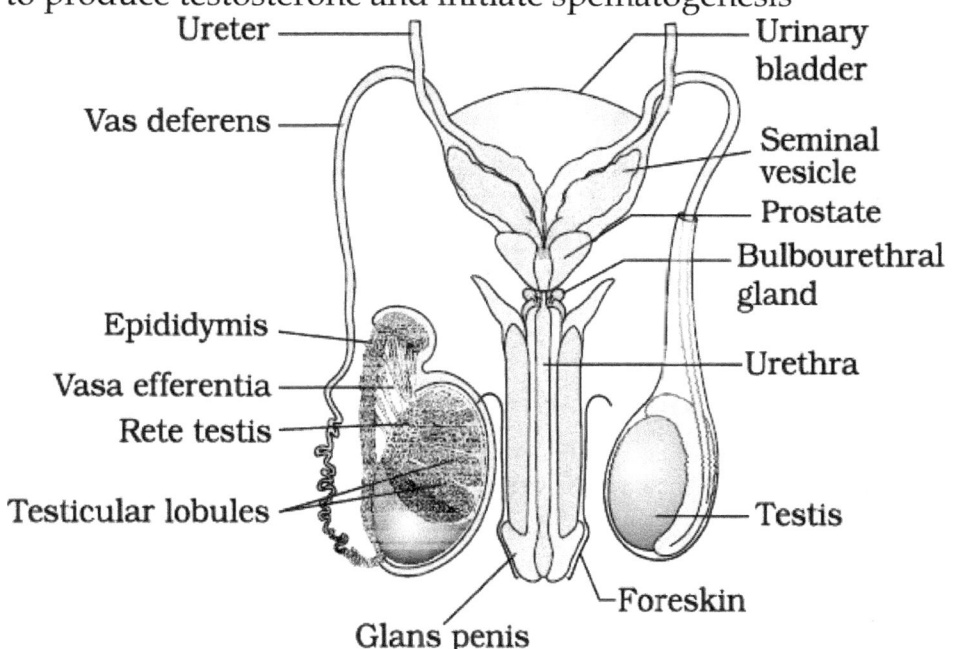

Female Reproductive System

The female reproductive system is a complex and intricately designed network of organs that work together to facilitate the processes of reproduction and hormonal regulation. Comprising both internal and external structures, this system plays a central role in the development of female sexual characteristics, the menstrual cycle, and the creation of new life.

Internal Structures:

i. **Ovaries**: Paired organs located on either side of the uterus, the ovaries are the primary reproductive organs in females. They produce eggs (ova) and release hormones, including estrogen and progesterone.
ii. **Fallopian Tubes:** Extending from the ovaries to the uterus, the fallopian tubes serve as conduits for the transportation of eggs. Fertilization typically occurs in the fallopian tubes when a sperm meets an egg.
iii. **Uterus:** Shaped like an inverted pear, the uterus is a muscular organ where a fertilized egg implants and grows into a fetus during pregnancy. The uterus undergoes contractions during labor to facilitate childbirth.
iv. **Cervix:** The lower part of the uterus, the cervix, acts as a gateway between the uterus and the vagina. It allows the passage of menstrual blood and serves as the point of entry for sperm during intercourse.
v. **Vagina:** The vagina is a muscular tube connecting the cervix to the external genitalia. It serves as the birth canal during delivery and also allows for the passage of menstrual blood and sexual intercourse.

External Structures

i. **Labia Majora and Labia Minora:** The labia majora are the outer folds of skin surrounding the vaginal opening, while the labia minora are the inner folds. These structures provide protection for the sensitive genital tissues.
ii. **Clitoris:** A small, highly sensitive organ located at the top of the vulva, the clitoris contains numerous nerve endings and plays a key role in sexual arousal.

iii. **Vulva:** The collective term for the external female genitalia, the vulva encompasses the labia, clitoris, vaginal opening, and urethral opening.

Hormonal Regulation: The female reproductive system is intricately regulated by hormones, primarily estrogen and progesterone. These hormones, produced by the ovaries, influence the menstrual cycle, support pregnancy, and play a vital role in maintaining female reproductive health.

Reproductive Processes

1. **Gametogenesis:** Gametogenesis is the process of producing specialized cells called gametes (sperm and eggs) through meiosis. In males, this process occurs in the testes, leading to the production of sperm. In females, it occurs in the ovaries, resulting in the formation of eggs.
2. **Fertilization:** Fertilization occurs when a sperm cell combines with an egg cell, resulting in the formation of a zygote. Fertilization usually takes place in the fallopian tubes.
3. **Pregnancy and Gestation:** In mammals, pregnancy occurs when a fertilized egg implants and develops into an embryo in the uterus. The embryo undergoes further development, known as gestation, leading to the birth of offspring.
4. **Menstrual Cycle:** In females, the menstrual cycle involves the regular shedding of the uterine lining (menstruation) and the preparation of the uterus for potential implantation of a fertilized egg.

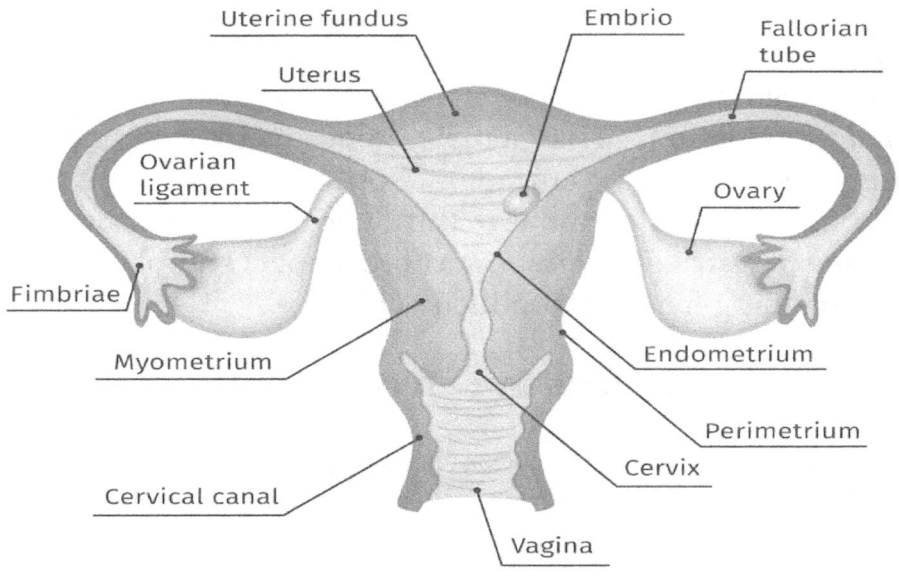

Endocrine System

The endocrine system is a complex network of glands and organs that secrete hormones to regulate various physiological processes in the body. It plays a crucial role in maintaining homeostasis, growth and development, metabolism, reproduction, and response to stress.

Glands Of The Endocrine System

1. **Pituitary Gland:** Often called the "master gland," the pituitary gland is located at the base of the brain and produces numerous hormones that control other endocrine glands.

2. **Thyroid Gland:** Situated in the neck, the thyroid gland produces hormones that regulate metabolism, growth, and development.
3. Parathyroid Glands: These small glands are located near the thyroid gland and secrete hormones that regulate calcium and phosphate levels in the body.
4. **Adrenal Glands:** The adrenal glands, situated on top of the kidneys, produce hormones involved in stress response (such as cortisol) and regulation of electrolyte balance (such as aldosterone).
5. **Pancreas:** The pancreas serves as both an exocrine and endocrine gland. It secretes digestive enzymes into the digestive tract and produces hormones like insulin and glucagon, which regulate blood glucose levels.
6. **Gonads:** In males, the testes produce testosterone, while in females, the ovaries produce estrogen and progesterone. These hormones are involved in reproductive functions and secondary sexual characteristics.
7. **Pineal Gland:** The pineal gland, located in the brain, produces the hormone melatonin, which regulates sleep-wake cycles and seasonal rhythms.

Hormones: Hormones are chemical messengers secreted by endocrine glands into the bloodstream. They travel throughout the body and bind to specific target cells or organs, initiating a response. Hormones can be classified into different categories, including peptide hormones (such as insulin), steroid hormones (such as cortisol), and amino acid-derived hormones (such as epinephrine). Each hormone has specific target cells or organs that possess receptors for that hormone. The binding of the hormone to its receptor triggers specific cellular responses and physiological effects.

Regulation and Feedback Mechanisms

Regulation and feedback mechanisms are integral components of biological systems, ensuring the dynamic balance and functionality of various physiological processes. In living organisms, homeostasis, or the maintenance of internal stability, is achieved through intricate regulatory mechanisms that respond to internal and external changes.Regulation and feedback mechanisms are crucial for maintaining the stability and functionality of living organisms. They allow organisms to adapt to changing environments, respond to stressors, and ensure that internal conditions remain within narrow ranges compatible with life. Without these mechanisms, the dynamic equilibrium required for optimal physiological function would be compromised

Biological regulation involves the control of a system's variables to maintain an optimal and stable internal environment. This is vital for the proper functioning of cells, tissues, and organs. The endocrine and nervous systems are primary regulators, orchestrating responses to stimuli and coordinating various physiological activities.Feedback mechanisms are essential elements of regulation, allowing the system to monitor its outputs and adjust its activities accordingly. There are two main types of feedback: positive and negative.

- **Negative Feedback:** This is the more common type of feedback in biological systems. It works to counteract changes and maintain stability. For example, in temperature regulation, if the body temperature rises, negative feedback mechanisms activate cooling responses like sweating to bring the temperature back to the set point.
- **Positive Feedback:** While less common, positive feedback amplifies changes in the system. This is often seen in processes that require rapid and decisive action. For instance, during childbirth, positive feedback mechanisms intensify

contractions until delivery occurs. Once the goal is achieved, the feedback loop is interrupted.

Examples Of Regulatory Systems

1. **Thermoregulation:** The body maintains a stable internal temperature through negative feedback. When temperature deviates from the set point, mechanisms like shivering or sweating are activated to restore equilibrium.
2. **Blood Glucose Regulation:** The endocrine system, particularly the pancreas, regulates blood glucose levels. When blood glucose rises after a meal, insulin is released to facilitate glucose uptake by cells, lowering blood sugar levels.
3. **Hormonal Regulation of the Menstrual Cycle:** The female reproductive system is regulated by hormonal feedback mechanisms. The interplay of hormones like estrogen and progesterone controls the menstrual cycle, ensuring proper ovulation and preparation for potential pregnancy.
4. **Neurotransmitter Regulation in Nervous System:** Neurons in the nervous system communicate through neurotransmitters. Feedback mechanisms regulate the release and reception of neurotransmitters, maintaining appropriate signaling between nerve cells.

Interactions with the Nervous System

The endocrine system and the nervous system often work together to coordinate and regulate bodily functions. The hypothalamus, located in the brain, serves as a link between the endocrine and nervous systems, controlling hormone secretion from the pituitary gland. The hypothalamus communicates with the pituitary gland through a complex network of nerves and releasing hormones, ensuring proper regulation of hormone production and release.

Enzymes	Location	Substrate	End Product
Amylase	Salivary gland, pancreas, small intestine	Starch	Maltose
Lipase	Pancreas, small intestine	Lipids	Fatty and glycerol
Protease	Stomach, Pancreas, small intestine	Proteins	Peptide and amino acid
Sucrase	Small intestine	Sucrose	Glucose and fructose
Lactase	Small intestine	Lactose	Glucose and galactose
Maltase	Small intestine	Maltose	Glucose
DNA Polymerase	Nucleus, mitochondria, chloroplasts	DNA template	Synthesis of complementary DNA strand
RNA Polymerase	Nucleus	DNA template	Synthesis of RNA
Catalase	Peroxisomes	Hydrogen peroxides	Water and Oxygen
Pepsin	Stomach	Proteins	Peptides
Trypsin	Pancreas, Small intestine	Proteins	Peptides and Amino acids
Cellulase	Microorganisms, digestive system	Cellulose	Glucose

ECOLOGY AND ECOSYSTEMS — CHAPTER 7

Introduction To Ecology And Ecosystem

Ecology is the scientific study of the relationships between organisms and their environment. It seeks to understand how living organisms interact with each other and with their physical surroundings. Ecology explores the distribution and abundance of organisms, their interactions, and the processes that shape their populations, communities, and ecosystems. By studying ecology, scientists can gain insights into the intricate web of life and the intricate balance that sustains ecosystems.

Types of Ecology

a. **Autecology:** Autecology focuses on studying individual organisms and their interactions with the environment. It examines how organisms adapt, survive, and reproduce in specific ecological conditions.

b. **Synecology:** Synecology, also known as community ecology, investigates the interactions between populations of different species that coexist in the same geographical area. It examines how species interact, compete, and depend on each other for resources.

c. **Population Ecology:** Population ecology studies the dynamics and characteristics of populations of a single species. It analyzes factors such as birth rates, death rates, dispersal patterns, and population growth to understand the factors that influence population size and structure.

d. **Behavioral Ecology:** Behavioral ecology explores the behavior of organisms in relation to their environment. It examines how behaviors such as foraging, mating, communication, and territoriality are shaped by ecological factors and natural selection.
e. **Conservation Ecology:** Conservation ecology focuses on understanding and preserving biodiversity and ecosystems. It addresses the impacts of human activities on the environment and develops strategies for the sustainable management and conservation of natural resources.

An ecosystem is a dynamic and interconnected system composed of living organisms and their physical environment. It encompasses the interactions between biotic (living) and abiotic (non-living) components in a specific area. Ecosystems range in size and complexity, from small ponds to vast forests or entire biomes. Ecosystems are characterized by energy flow and the cycling of nutrients, as well as the interactions between organisms and their physical surroundings.

Ecosystems consist of:
i. **Producers:** Organisms such as plants or algae that convert sunlight into chemical energy through photosynthesis.
ii. **Consumers:** Organisms that obtain energy by consuming other organisms, including herbivores (plant-eaters), carnivores (meat-eaters), and omnivores (consumers of both plants and animals).
iii. **Decomposers:** Organisms like bacteria and fungi that break down dead organic matter, releasing nutrients back into the ecosystem.
iv. **Abiotic Factors:** Non-living components of the ecosystem, including temperature, water availability, soil composition, sunlight, and climate.

Introduction To Biology

■ Levels Of Organization In Ecology

Ecology studies the interactions and relationships between organisms and their environment. These interactions occur at different levels of organization, each contributing to the understanding of ecological systems. Here are the levels of organization in ecology:

1. **Organism:** The individual living entity is the basic unit of ecology. It refers to a single organism, such as a plant, animal, or microorganism. Studying the adaptations, behaviors, and physiology of organisms provides insights into their survival strategies and ecological roles.

2. **Population:** A population consists of a group of individuals of the same species living in the same area at the same time. Population ecology focuses on studying the dynamics of populations, including population size, density, growth rates, and factors that influence population changes, such as birth rates, death rates, and migration.

3. **Community:** A community is an assemblage of populations of different species that coexist and interact in the same geographic area. Community ecology examines the interactions, relationships, and species diversity within a community. It explores concepts such as competition, predation, symbiosis, and the effects of species interactions on community structure and stability.

4. **Ecosystem:** An ecosystem encompasses both the biotic (living) and abiotic (non-living) components in a particular area. It includes the interactions between organisms and their physical environment. Ecosystem ecology studies the flow of energy, cycling of nutrients, and the processes that shape the structure and functioning of ecosystems.

5. **Landscape:** A landscape refers to a larger geographic area consisting of multiple interacting ecosystems. It involves the study of spatial patterns, connectivity, and the flow of energy

and resources across different ecosystems. Landscape ecology explores how landscape features and habitat fragmentation influence ecological processes and species distribution.
6. **Biome:** A biome is a large-scale ecological region characterized by distinctive climate, vegetation, and animal communities. Biomes are influenced by factors such as temperature, precipitation, and soil conditions. Examples of biomes include tropical rainforests, deserts, grasslands, and tundra. Biogeography focuses on the distribution of biomes and the factors that shape their boundaries.
7. **Biosphere:** The biosphere is the global ecological system that encompasses all the Earth's ecosystems. It includes the sum of all living organisms, their interactions, and the physical environments they inhabit. The study of the biosphere involves understanding the global processes and interconnectedness of ecosystems on Earth.

■ Ecosystems And Biomes

Ecosystems and biomes are fundamental concepts in ecology that provide insight into the structure, function, and distribution of ecological systems on Earth. They help us understand the intricate relationships between living organisms and their physical environment, shedding light on the complexity and interconnectedness of nature. Let's delve into a detailed and comprehensive explanation of ecosystems and biomes.

Ecosystems

An ecosystem is a dynamic and self-contained system comprising both living organisms and their physical environment. It encompasses the interactions and exchanges of energy, nutrients, and materials among organisms and their surroundings. Ecosystems can vary in size and complexity, ranging from a small pond or forest to an entire biome.

Components of Ecosystems

i. **Biotic Components:** These are the living organisms within an ecosystem, including plants, animals, microorganisms, and decomposers. Biotic components interact with one another, forming intricate webs of relationships and dependencies.
ii. **Abiotic Components:** These are the non-living factors that influence the functioning of an ecosystem. They include physical factors such as sunlight, temperature, water availability, soil composition, and nutrients. Abiotic components provide the necessary conditions for life and shape the characteristics of the biotic community.

Types of Ecosystems

Ecosystems can be classified into various types based on their predominant features and characteristics:

1. **Terrestrial Ecosystems:** These include forests, grasslands, deserts, tundra, and other land-based ecosystems. They are shaped by factors such as climate, topography, and soil composition.
2. **Aquatic Ecosystems:** These encompass freshwater ecosystems (such as rivers, lakes, and wetlands) and marine ecosystems (such as oceans, coral reefs, and estuaries). Aquatic ecosystems are influenced by factors such as water depth, flow rate, salinity, and nutrient availability.
3. **Anthropogenic Ecosystems:** These are human-created or modified ecosystems, such as agricultural lands, urban areas, and managed forests. Anthropogenic ecosystems have distinct characteristics and are influenced by human activities.

Introduction To Biology

Biomes

Biomes are large-scale ecological regions characterized by distinct climate, vegetation, and animal communities. They represent major terrestrial or aquatic ecosystems found across different continents. Biomes are primarily determined by climatic factors, including temperature, precipitation, and seasonality. They provide a broad framework for understanding global patterns of biodiversity and ecosystem distribution.

Major Terrestrial Biomes

Tropical Rainforest

Tropical rainforests are incredibly diverse and complex ecosystems found near the equator. They are characterized by their lush vegetation, high rainfall, and warm temperatures throughout the year. Here are some key features of tropical rainforests:

i. **Biodiversity:** Tropical rainforests are the most biologically diverse ecosystems on Earth, harboring a wide variety of plant and animal species. They contain an extraordinary abundance of species, including many endemic and unique organisms. The dense vegetation provides a multitude of niches and habitats, supporting an intricate web of interactions among species.

ii. **Dense Vegetation:** Tropical rainforests are known for their dense and multi-layered vegetation. They have a complex structure with emergent trees towering above the canopy layer, an upper canopy layer receiving the most sunlight, an understory layer with smaller trees and shrubs, and a forest floor covered with decomposing organic matter. This layered structure provides a range of microhabitats and creates a high level of competition for sunlight among plants.

iii. **Climate:** The climate of tropical rainforests is characterized by high temperatures and humidity. They typically receive heavy rainfall throughout the year, often exceeding 2,000 millimeters (80 inches) annually. The consistent warm temperatures and abundant moisture create an ideal environment for plant growth and support the lush vegetation.
iv. **Rich Soil:** Despite the immense productivity of tropical rainforests, the soil is generally nutrient-poor. Most of the nutrients are quickly absorbed and recycled within the ecosystem due to the rapid decomposition of organic matter. The soil is typically acidic and shallow, with nutrients being concentrated in the biomass of plants rather than the soil itself.
v. **Canopy Interactions:** The dense canopy of tropical rainforests creates a unique environment where competition for sunlight is intense. Plants have adapted various strategies to access light, such as growing tall, having large leaves, or climbing on the trunks of trees. Epiphytic plants, including orchids and bromeliads, grow on branches and tree trunks to reach the light. This complex vertical structure fosters a multitude of interactions among plants, including competition, cooperation, and epiphyte-host relationships.
vi. **Fauna:** Tropical rainforests support an incredible diversity of animal species. They are home to numerous mammals, birds, reptiles, amphibians, and insects. Many species have developed specialized adaptations, such as camouflage, elaborate mating displays, and unique feeding habits. Examples of iconic rainforest animals include jaguars, toucans, monkeys, tree frogs, and colorful butterflies.

Unfortunately, tropical rainforests face significant threats due to deforestation, habitat fragmentation, and climate change. Human activities such as logging, agriculture (especially for palm oil and cattle ranching), and infrastructure development have led to substantial loss of rainforest areas. Conservation efforts aim to protect and restore these valuable ecosystems, recognizing their crucial role in maintaining global biodiversity, regulating climate, and providing ecosystem services.

Temperate Deciduous Forest

These biomes are found in regions with moderate climates, distinct seasons, and deciduous trees that shed their leaves in the winter. Temperate deciduous forests are characterized by specific features that distinguish them from other biomes.

i. **Climate:** Temperate deciduous forests experience moderate climates with four distinct seasons - spring, summer, autumn, and winter. They are found in regions with moderate rainfall evenly distributed throughout the year.

ii. **Vegetation:** The dominant vegetation consists of deciduous trees, which shed their leaves during the winter months. Common tree species in temperate deciduous forests include oak, maple, beech, birch, and hickory. These trees have broad leaves that capture sunlight during the growing season and provide vibrant colors in the autumn.

iii. **Biodiversity:** Temperate deciduous forests are known for their rich biodiversity. They support a wide range of plant and animal species, including various types of mammals, birds, reptiles, amphibians, and insects. The diverse canopy structure and the presence of different vegetation layers create microhabitats for numerous organisms.

iv. **Stratification of Vegetation:** The vegetation in temperate deciduous forests is typically stratified into distinct layers. The tallest trees form the upper canopy layer, which receives the most sunlight. Below the canopy, there is an understory layer consisting of smaller trees, shrubs, and saplings. Finally, the forest floor is covered with herbaceous plants, mosses, ferns, and leaf litter.

v. **Leaf Seasonality:** One of the defining features of temperate deciduous forests is leaf seasonality. In spring, the trees leaf out, creating a dense and vibrant canopy. During the autumn, the leaves change color, displaying a stunning array of red, orange, and gold before eventually falling off in preparation for winter.

vi. **Soil Composition:** The soils in temperate deciduous forests tend to be fertile and well-drained. The leaf litter that accumulates on the forest floor decomposes and enriches the soil with organic matter, supporting the growth of diverse plant species.

vii. **Wildlife Adaptations:** The animal species in temperate deciduous forests have various adaptations to survive the changing seasons. Migratory birds and some mammals, such as bears and squirrels, store food during the abundant seasons to sustain themselves through the winter months. Some animal species hibernate, while others have specific adaptations for coping with cold temperatures and scarcity of resources.

viii. **Human Influence:** Temperate deciduous forests have experienced significant human impact through deforestation, land conversion for agriculture, and urbanization. However, efforts are being made to conserve and restore these forests due to their ecological importance and biodiversity.

Temperate deciduous forests provide essential ecosystem services, including carbon sequestration, water regulation, soil stability, and habitat for numerous species. They are also valued for recreational activities, such as hiking, camping, and wildlife observation. Understanding the features of temperate deciduous forests helps us appreciate their beauty, biodiversity, and ecological significance.

Taiga (Boreal Forest)

The Taiga, also known as the Boreal Forest or Coniferous Forest, is a biome characterized by unique features that enable it to thrive in cold and harsh environments, located in northern regions, these biomes have cold winters, coniferous forests, and a low diversity of plant and animal species.

i. **Climate:** The Taiga experiences long, cold winters and short, cool summers. It is located in high-latitude regions, mostly in the Northern Hemisphere, such as Canada, Russia, and Scandinavia. Average temperatures range from -40°C (-40°F) in winter to 20°C (68°F) in summer, with annual precipitation ranging from 40-100 cm (16-39 inches).

ii. **Vegetation:** The dominant vegetation in the Taiga biome consists of coniferous evergreen trees, such as spruce, fir, pine, and larch. These trees are adapted to the cold climate and have needle-like leaves that minimize water loss. The conical shape of the trees helps shed snow and prevents branches from breaking under heavy snowfall.

iii. **Biodiversity:** Although the Taiga is predominantly characterized by coniferous trees, it supports a diverse range of plant and animal species. Common plant species found in the Taiga include mosses, lichens, and dwarf shrubs. Animal species that inhabit the Taiga include mammals such as moose, reindeer, wolves, bears, and smaller mammals like squirrels, lynx, and ermines. Birds like owls, woodpeckers, and various migratory birds are also found in this biome.

iv. **Adaptations:** Both plant and animal species in the Taiga have evolved various adaptations to survive in the extreme conditions. Coniferous trees have needle-like leaves that reduce water loss and can photosynthesize even in low temperatures. Some animals, like the Arctic fox and snowshoe hare, change their fur color from brown in summer to white in winter for camouflage. Other animals hibernate or migrate during winter to cope with the scarcity of food.

v. **Soil:** The Taiga has nutrient-poor soils, known as podzols or spodosols, which are acidic and have a high concentration of organic matter. The cold climate and slow decomposition of organic matter lead to the accumulation of thick layers of leaf litter and peat.

vi. **Threats and Conservation:** The Taiga biome faces threats from deforestation, industrial activities, and climate change. Logging for timber, mining, and oil and gas extraction pose significant challenges to the preservation of this delicate ecosystem. Conservation efforts focus on sustainable forestry practices, protected areas, and raising awareness about the importance of the Taiga's biodiversity and ecosystem services.

The Taiga biome is an important part of the Earth's natural heritage. Its unique features and adaptations allow for the survival and coexistence of a diverse range of plants and animals in challenging environmental conditions. Understanding and conserving the Taiga biome are crucial for maintaining biodiversity, mitigating climate change, and preserving one of the largest terrestrial biomes on the planet.

Grasslands

Grasslands are vast terrestrial ecosystems characterized by grasses as the dominant vegetation type, with a limited number of trees or shrubs. They are found in both tropical and temperate regions around the world. Grasslands exhibit distinct features that contribute to their unique ecological characteristics.

i. **Vegetation:** Grasslands are primarily composed of various species of grasses, which are well-adapted to the semi-arid or sub-humid climate conditions. They have deep root systems that allow them to withstand drought and fire. However, grasslands can also contain scattered shrubs, herbs, and flowering plants, which provide additional diversity to the ecosystem.

ii. **Climate:** Grasslands experience a range of climates, including hot summers and cold winters in temperate regions and year-round warmth in tropical regions. They often receive moderate rainfall, but it is often seasonal and subject to periodic droughts. These climate conditions favor the growth of grasses over other vegetation types.

iii. **Fire Adaptation:** Grasslands have evolved to tolerate and even benefit from periodic fires. Fire helps maintain the open nature of the grasslands by preventing the encroachment of woody vegetation, stimulating the growth of new grass shoots, and cycling nutrients. Grasses have adaptations such as underground stems (rhizomes) and buds (tillers) that allow them to regrow after fire events.

iv. **Biodiversity:** Although grasslands may appear visually uniform, they can support a surprisingly diverse array of plant and animal species. Grasslands are home to a variety of herbivores, such as bison, antelope, gazelles, and zebras, which rely on the abundance of grass for grazing. These herbivores, in turn, attract predators such as lions, cheetahs, wolves, and birds of prey. Additionally, grasslands often

provide important habitats for insects, reptiles, birds, and small mammals.
v. **Adapted Grazers:** The structure of grasslands, with their expansive grass cover, makes them ideal habitats for grazing animals. Grazers have specialized digestive systems and teeth adapted for efficiently consuming grass. They have evolved to coexist with the grasses, often moving in herds to maximize foraging opportunities and minimize predation risk.
vi. **Nutrient Cycling:** Grasslands have nutrient-rich soils due to the rapid decomposition of grasses and the cycling of nutrients within the ecosystem. Grazing animals play a vital role in nutrient cycling by consuming vegetation, depositing waste (feces), and redistributing nutrients through their movements.
vii. **Threats and Conservation:** Grasslands face various threats, including habitat loss due to conversion to agriculture, urban development, and overgrazing by domestic livestock. Climate change can also impact grasslands, affecting rainfall patterns and contributing to desertification in some regions. Conservation efforts aim to protect and restore grassland ecosystems, promoting sustainable land management practices and preserving the unique biodiversity they support.

Introduction To Biology

Desert

Deserts are unique and extreme ecosystems characterized by specific features and adaptations to harsh environmental conditions.

i. **Aridity:** Deserts are known for their extreme dryness, with very low precipitation levels. They receive less than 250 millimeters (10 inches) of rainfall per year, and some deserts experience even less. The aridity of deserts is often due to factors such as geographic location, atmospheric conditions, and the presence of mountain ranges that block moisture-laden winds.

ii. **High Temperature Extremes:** Deserts are characterized by significant temperature variations. They can experience scorching daytime temperatures, reaching well above 40 degrees Celsius (104 degrees Fahrenheit), and cold nighttime temperatures that can drop below freezing. The lack of cloud cover and humidity contributes to these temperature extremes.

iii. **Sparse Vegetation:** Vegetation in deserts is generally sparse and adapted to survive with minimal water availability. Plants have evolved various strategies to cope with the arid conditions, such as having deep root systems to access groundwater, storing water in fleshy stems or leaves, or reducing leaf surface area to minimize water loss through transpiration. Common desert plants include cacti, succulents, and shrubs.
iv. **Adapted Animal Life:** Desert animals have evolved unique adaptations to survive in the harsh desert environment. They often have physiological and behavioral adaptations that enable them to conserve water, tolerate high temperatures, and obtain water and nutrients from scarce food sources. Examples of desert animals include camels, snakes, lizards, scorpions, and rodents.
v. **Sand Dunes and Rock Formations:** Deserts are often characterized by impressive sand dunes formed by wind-driven sand particles. These dunes can have different shapes and sizes, ranging from crescent-shaped barchan dunes to long linear ridges called longitudinal dunes. Deserts also feature fascinating rock formations, such as mesas, buttes, and canyons, shaped by erosion over millions of years.
vi. **Limited Surface Water:** Surface water sources in deserts are rare and ephemeral, appearing temporarily after rain events. Deserts may have intermittent rivers or arroyos, dry riverbeds that occasionally fill with water during heavy rainfall. Oasis ecosystems, where groundwater reaches the surface, provide essential water sources and support unique biodiversity hotspots.
vii. **Extreme Isolation:** Deserts often have vast expanses of uninhabited and isolated areas, creating unique ecological niches and limited human presence. This isolation can result

in distinct evolutionary patterns and high levels of endemism, where species are found only in specific desert regions.

Despite the harsh conditions, deserts exhibit surprising ecological diversity and resilience. They have adapted to extreme aridity and are home to a wide array of unique plant and animal species. Deserts also provide important ecological services, such as water and nutrient cycling, and have cultural significance for indigenous communities who have thrived in these environments for centuries.

Tundra

The tundra is a unique and extreme biome characterized by its cold climate, low temperatures, short growing season, and limited vegetation.

i. **Climate:** The tundra experiences long, cold winters and short, cool summers. Average temperatures in the tundra rarely rise above freezing, and the ground remains frozen for most of the year (permafrost). The annual precipitation is low, primarily in the form of snow, and strong winds are common.

ii. **Vegetation:** Tundra vegetation is adapted to survive in the harsh conditions of low temperatures and short growing seasons. The dominant plant life includes low-growing shrubs, mosses, lichens, and small herbaceous plants. Trees are generally absent in the tundra due to the cold temperatures and shallow permafrost.
iii. **Permafrost:** Permafrost is a permanently frozen layer of soil that characterizes the tundra. It restricts the depth of plant root systems and affects water drainage in the ecosystem. Permafrost also acts as a barrier, preventing water from penetrating deep into the ground.
iv. **Biodiversity:** Despite the challenging conditions, the tundra supports a surprising variety of wildlife. Animal species in the tundra include migratory birds, such as geese and ducks, small mammals like lemmings and Arctic hares, caribou, muskoxen, wolves, and polar bears. Many of these animals have adaptations to cope with the cold climate, such as thick fur, layers of fat, and specialized behaviors.
v. **Adaptations:** Tundra plants and animals have unique adaptations to survive in the extreme environment. Examples include small, compact growth forms to minimize exposure to the cold and wind, deep roots to access limited water resources, and thick fur or feathers for insulation. Animals often have behavioral adaptations like hibernation or migration to find food and avoid harsh winter conditions.
vi. **Fragile Ecosystem:** The tundra ecosystem is delicate and vulnerable to disturbances. Due to the short growing season, vegetation recovery from disturbances such as human activities or natural events can take decades or even centuries. Human activities, including oil and gas extraction, mining, and climate change, pose significant threats to the tundra biome and its fragile balance.

vii. **Role in Climate Regulation:** The tundra plays an essential role in global climate regulation. The vast areas of permafrost store significant amounts of carbon, which, if released due to permafrost thawing, can contribute to increased greenhouse gas emissions and further climate change.
viii. **Limited Human Presence:** The tundra has a sparse human population due to its harsh conditions. Indigenous communities have adapted their lifestyles to the tundra environment, relying on traditional activities such as hunting, herding, and fishing.

The tundra's unique characteristics and biodiversity make it a valuable ecosystem to study and preserve. Understanding its adaptations and vulnerability helps in conservation efforts and the sustainable management of this fragile biome.

Major Aquatic Biomes

Freshwater

Freshwater ecosystems are aquatic environments that contain relatively low salt concentrations, making them distinct from marine ecosystems. These ecosystems include rivers, lakes, ponds, wetlands, and streams, and they play a vital role in supporting a wide variety of plant and animal species.

i. **Water Source:** Freshwater ecosystems are primarily sustained by freshwater sources such as rainfall, melting snow, and underground springs. These sources provide a continuous supply of water that supports the ecosystem's functioning.

ii. **Flowing and Standing Water:** Freshwater ecosystems can be categorized into flowing water systems, such as rivers and streams, and standing water systems, such as lakes, ponds, and wetlands. Flowing water systems have a continuous movement of water, while standing water systems are relatively stagnant.

iii. **Nutrient Availability:** Freshwater ecosystems receive nutrients from various sources, including weathering of rocks, decaying organic matter, and inputs from surrounding terrestrial ecosystems. These nutrients, such as nitrogen and phosphorus, are essential for the growth and productivity of aquatic plants and algae.

iv. **Light Penetration:** Sunlight plays a critical role in freshwater ecosystems as it provides energy for photosynthesis. The penetration of light into the water column varies depending on factors such as water depth, water clarity, and the presence of suspended particles or algae.

v. **Biodiversity:** Freshwater ecosystems support a high level of biodiversity. They are home to numerous species of fish, amphibians, reptiles, invertebrates, and plants. The diverse range of habitats within freshwater ecosystems, such as riffles, pools, wetlands, and submerged vegetation, provides niches for different organisms.

vi. **Oxygen Levels:** Oxygen availability is crucial for the survival of aquatic organisms. In freshwater ecosystems, oxygen dissolves into the water from the atmosphere and through the process of photosynthesis by aquatic plants. Adequate oxygen levels are essential for the respiration of aquatic organisms.

vii. **Nutrient Cycling:** Freshwater ecosystems have nutrient cycles that involve the uptake, utilization, and recycling of nutrients by various organisms. Nutrient cycling is vital for the growth of aquatic plants, algae, and other primary producers, which form the base of the food web.

viii. **Habitat Heterogeneity:** Freshwater ecosystems exhibit a range of habitats, including shallow areas, deep zones, rocky substrates, sandy bottoms, and vegetated areas. This habitat heterogeneity provides diverse niches and shelter for different species, contributing to overall ecosystem productivity and biodiversity.

ix. **Human Impact:** Freshwater ecosystems face various threats from human activities, such as pollution, habitat destruction, water extraction, and the introduction of non-native species. These impacts can disrupt the delicate balance of the ecosystem and have significant consequences for the organisms and ecological processes within.

Marine

Marine ecosystems, also known as the marine biome, encompass all the bodies of saltwater on Earth, including oceans, seas, coral reefs, estuaries, and coastal areas. They are vast and diverse, covering about 70% of the planet's surface. Here are some key features of marine ecosystems:

i. **Salinity:** Marine ecosystems have high salinity due to the presence of dissolved salts in seawater. The average salinity of the oceans is about 3.5%, which means that for every 1,000 grams of seawater, around 35 grams are salts.

ii. **Temperature and Light:** Marine ecosystems experience a wide range of temperatures, from polar regions with freezing temperatures to tropical regions with warm waters. Temperature variations influence the distribution of marine organisms. Light availability decreases with depth, affecting the types of organisms found at different depths in the water column.

iii. **Biodiversity:** Marine ecosystems are incredibly diverse and support a wide array of plant and animal species. They are home to countless marine organisms, including phytoplankton, zooplankton, fish, marine mammals, reptiles, invertebrates, and coral reefs. The biodiversity of marine ecosystems is crucial for maintaining ecosystem balance and providing numerous ecosystem services.

iv. **Productivity:** Marine ecosystems are highly productive, with primary production driven by phytoplankton (microscopic plants) that use sunlight and nutrients to convert carbon dioxide into organic matter through photosynthesis. This primary production forms the base of the marine food web, sustaining the entire ecosystem.

v. **Physical Features:** Marine ecosystems consist of various physical features, including continental shelves, abyssal plains, trenches, seamounts, and underwater volcanic structures. These physical features create diverse habitats and niches for marine organisms, contributing to the overall biodiversity of the marine biome.

vi. **Coral Reefs:** Coral reefs are unique and highly productive marine ecosystems. They are formed by colonies of tiny animals called coral polyps that secrete calcium carbonate skeletons. Coral reefs are home to a vast array of marine species and provide critical habitats, protect coastlines from erosion, and support tourism and fisheries industries.

vii. **Ocean Currents:** Marine ecosystems are influenced by ocean currents, which are driven by wind patterns, temperature differences, and the Earth's rotation. Ocean currents play a vital role in redistributing heat, nutrients, and organisms, shaping marine ecosystems' characteristics and influencing global climate patterns.

viii. **Threats and Conservation:** Marine ecosystems face numerous threats, including overfishing, pollution, habitat destruction, climate change, and ocean acidification. These threats can have detrimental effects on marine biodiversity and ecosystem health. Conservation efforts aim to protect and sustainably manage marine ecosystems, including the establishment of marine protected areas, sustainable fishing practices, and reducing pollution inputs.

■ Energy Flow In Ecosystems

Energy flow is a fundamental process in ecosystems that describes how energy is transferred and transformed as it moves through different organisms and trophic levels. It illustrates the flow of energy from the primary producers to consumers and decomposers, highlighting the interconnectedness and interdependence of organisms within an ecosystem. Let's explore the concept of energy flow in ecosystems in more detail.

1. **Autotrophs (Primary Producers):** Autotrophs, also known as primary producers, are organisms capable of synthesizing organic compounds from inorganic sources. They are typically photosynthetic organisms, such as plants, algae, and some

bacteria. These organisms capture energy from sunlight and convert it into chemical energy through the process of photosynthesis. In this process, they use carbon dioxide, water, and sunlight to produce glucose (a form of stored energy) and oxygen as a byproduct.
2. **Heterotrophs (Consumers):** Heterotrophs are organisms that obtain energy by consuming other organisms. They cannot produce their own organic compounds and rely on the energy stored in autotrophs and other heterotrophs. There are three main types of consumers:
3. **Herbivores:** Herbivores are primary consumers that feed exclusively on autotrophs, mainly plants or algae.
a. **Carnivores:** Carnivores are secondary and tertiary consumers that feed on other heterotrophs. They consume herbivores or other carnivores.
b. **Omnivores:** Omnivores are consumers that have a varied diet, feeding on both plants and animals.
4. **Decomposers:** Decomposers play a vital role in energy flow by breaking down organic matter and returning nutrients to the ecosystem. They include bacteria and fungi that decompose dead organisms, detritus (organic debris), and waste materials. Decomposers release nutrients back into the environment, allowing them to be recycled and utilized by primary producers once again.

Energy Flow and Trophic Levels

Energy flow in ecosystems occurs through trophic levels, which represent different feeding positions in a food chain or food web. Each trophic level has a specific energy relationship with the level above and below it.

1. **Primary Producers (Autotrophs):** Primary producers occupy the first trophic level and capture energy from the sun to produce organic compounds through photosynthesis.
2. **Primary Consumers (Herbivores):** Primary consumers occupy the second trophic level and feed directly on primary producers (autotrophs) for energy.
3. **Secondary Consumers (Carnivores):** Secondary consumers occupy the third trophic level and feed on primary consumers (herbivores) for energy.
4. **Tertiary Consumers (Carnivores):** Tertiary consumers occupy the fourth trophic level and feed on secondary consumers (carnivores or other consumers) for energy.
5. **Decomposers:** Decomposers occupy a separate trophic level as they obtain energy by breaking down organic matter. They contribute to nutrient recycling rather than being a part of the traditional food chain.

Energy Transfer and Energy Loss

Energy transfer from one trophic level to another is not 100% efficient. As energy flows through the ecosystem, there are energy losses in the form of heat, metabolic processes, and waste products. The amount of energy transferred from one trophic level to the next is typically around 10% of the energy available in the previous trophic level. This energy loss is due to metabolic activities, heat production, and incomplete digestion and assimilation of food.

Thermodynamics and Ecosystem

The laws of thermodynamics provide a fundamental understanding of energy transfer and energy loss in ecosystems.

In particular, the first and second laws of thermodynamics are relevant to the flow of energy in ecosystems.

First Law of Thermodynamics (Law of Conservation of Energy):
The first law of thermodynamics states that energy cannot be created or destroyed; it can only be transferred or converted from one form to another.

This law applies to energy flow in ecosystems, where energy is constantly transferred between organisms and the environment but is never created or destroyed, the first law of thermodynamics implies that the total amount of energy within a closed system remains constant. Energy enters an ecosystem through the primary producers (autotrophs) who capture solar energy through photosynthesis. This energy is then transferred to consumers and decomposers as they feed on other organisms. However, throughout this transfer, some energy is inevitably lost in the form of heat, metabolic processes, and waste products.

Second Law of Thermodynamics (Law of Entropy):
The second law of thermodynamics states that in any energy transfer or conversion, there is always an increase in entropy (disorder) in the universe.

This law has implications for energy flow in ecosystems, particularly regarding the efficiency of energy transfer and the quality of energy available at different trophic levels. As energy flows through an ecosystem, it undergoes transformations and transfers from one organism to another. At each transfer, some energy is lost as heat, which contributes to the increase in entropy. This loss of energy limits the efficiency of energy transfer between trophic levels in an ecosystem.

The second law of thermodynamics also highlights the concept of energy quality. Energy is most concentrated and useful in its initial form, such as sunlight captured by primary producers. However, as energy is transferred from one organism to another, its quality decreases. For example, energy stored in the chemical bonds of organic compounds in plants becomes less available and less concentrated as it is passed on to herbivores, carnivores, and decomposers.

The laws of thermodynamics provide a scientific basis for understanding the principles and limitations of energy transfer and energy loss in ecosystems. They emphasize that energy is conserved but transformed and degraded during its passage through the ecosystem, with some energy being lost as heat and increasing the overall disorder. These laws help us appreciate the complexity of energy flow in ecosystems and the challenges associated with sustaining and managing ecological systems.

Energy Pyramids

Energy pyramids illustrate the energy flow and biomass distribution in an ecosystem. They depict the decreasing energy available at each successive trophic level, with the primary producers forming the broad base and the higher trophic levels forming the narrower sections. The energy pyramid highlights the limited energy available to support organisms at higher trophic levels.

■ Nutrient Cycling

Nutrient cycling, also known as biogeochemical cycling, is the process by which nutrients are recycled and exchanged among living organisms and their environment in an ecosystem. It involves the continuous movement of essential elements, such as carbon, nitrogen, phosphorus, and other minerals, between the biotic (living) and abiotic (non-living) components of the ecosystem. Nutrient cycling plays a crucial role in maintaining the balance and sustainability of ecosystems. Let's explore the key aspects of nutrient cycling:

Nutrients enter ecosystems from both external and internal sources. External sources include the atmosphere, where gases like carbon dioxide and nitrogen can be taken up by plants, and weathering of rocks that releases minerals into the soil. Internal sources refer to the recycling of nutrients within the ecosystem through the decomposition of organic matter, waste materials, and dead organisms.

Nutrient Pathways

Nutrients follow specific pathways as they cycle through the ecosystem. The major pathways include:

a. **Absorption and Uptake:** Plants absorb nutrients from the soil or directly from the atmosphere through their leaves. They incorporate these nutrients into their tissues.

b. **Consumption:** Consumers, including herbivores, carnivores, and decomposers, obtain nutrients by consuming plants or other organisms.

c. **Decomposition:** Decomposers, such as bacteria and fungi, break down organic matter into simpler compounds, releasing nutrients back into the soil or water.

d. **Mineralization:** During decomposition, organic matter is converted into inorganic forms, such as ammonium (NH_4^+) or nitrate (NO_3^-), which can be taken up by plants.

e. **Assimilation:** Nutrients taken up by plants and consumed by animals are incorporated into their tissues and used for growth, development, and various metabolic processes.

f. **Release:** When organisms die or excrete waste materials, nutrients are returned to the environment, either directly or through decomposition.

Nutrient Cycling Processes

There are specific processes involved in nutrient cycling:
The carbon cycle is a biogeochemical cycle that describes the movement of carbon atoms between the atmosphere, land, oceans, and living organisms. It involves various processes, both natural and human-induced, which play a crucial role in regulating the Earth's climate. Here is an explanation of the carbon cycle:

Carbon Cycle

Carbon Sources:

Carbon enters the atmosphere through several sources:

i. **Respiration:** All living organisms, including plants, animals, and microorganisms, release carbon dioxide (CO_2) into the atmosphere through respiration.
ii. **Combustion:** The burning of fossil fuels, such as coal, oil, and natural gas, releases carbon dioxide into the atmosphere. This process is primarily associated with human activities.
iii. **Decomposition:** When dead plants and animals decay, carbon is released into the atmosphere as CO_2 through the process of decomposition.

Carbon Sink:

The oceans, forests, and other terrestrial ecosystems act as carbon sinks, absorbing and storing carbon dioxide from the atmosphere. This process is known as carbon sequestration.

i. **Ocean Absorption:** The oceans play a crucial role in absorbing atmospheric carbon dioxide through physical and chemical processes. Dissolved carbon dioxide forms carbonic acid, which leads to ocean acidification.

ii. **Photosynthesis:** Plants and other photosynthetic organisms take up carbon dioxide from the atmosphere and convert it into organic compounds through photosynthesis. This process removes carbon dioxide from the air and stores carbon in plant biomass.

Carbon Flux

The carbon flux refers to the movement of carbon between different reservoirs or sinks:

Biological Processes: Through the process of photosynthesis, plants convert atmospheric carbon dioxide into carbohydrates and other organic compounds. Carbon is then transferred through the food chain as organisms consume plants or other animals.

i. **Decomposition:** When plants or animals die, their organic matter is decomposed by bacteria and fungi. This releases carbon dioxide back into the atmosphere and completes the cycle.

ii. **Carbon Storage:** Carbon can be stored for long periods in the form of fossil fuels, such as coal, oil, and natural gas. These reservoirs represent ancient carbon that has been buried underground for millions of years.

iii. **Human Activities:** The burning of fossil fuels for energy, deforestation, and land-use changes contribute to the release of additional carbon dioxide into the atmosphere, disrupting the natural carbon cycle.

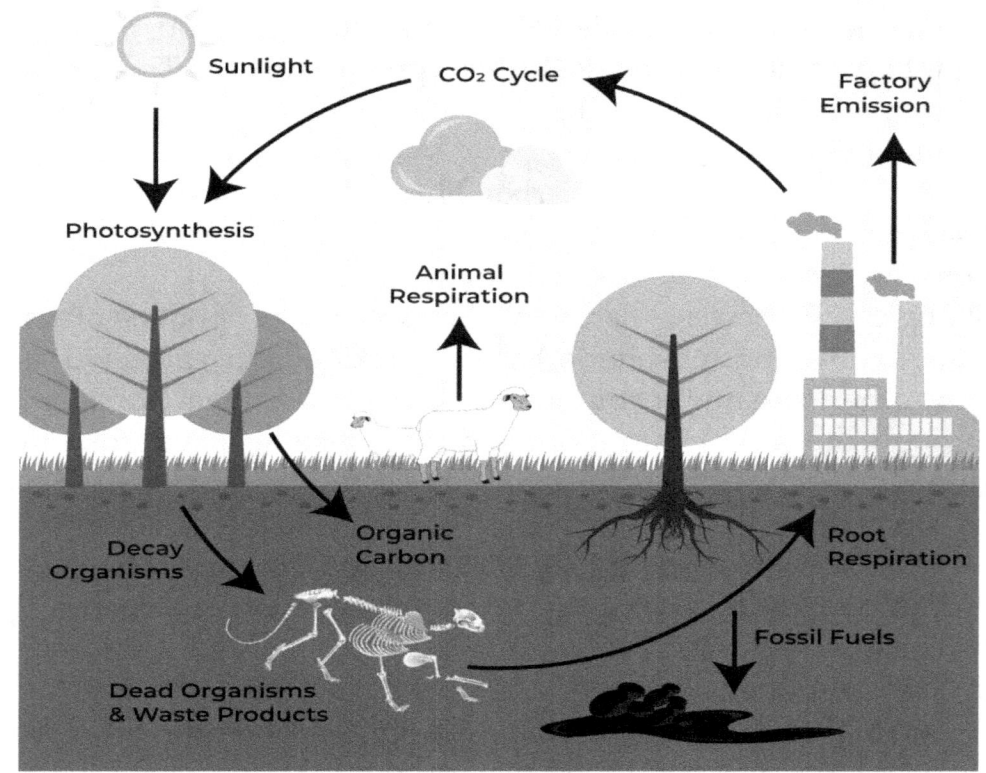

Nitrogen Cycle

The nitrogen cycle is a biogeochemical cycle that describes the transformation and circulation of nitrogen in the environment. Nitrogen is an essential element for the growth and development of living organisms, particularly for the formation of proteins, DNA, and other vital molecules. The nitrogen cycle involves various processes that convert nitrogen between different forms, making it available to different organisms.

i. **Nitrogen Fixation:** Nitrogen gas (N_2) makes up about 78% of the Earth's atmosphere, but most organisms cannot use nitrogen in this form. Nitrogen fixation is the process by which certain bacteria, known as nitrogen-fixing bacteria, convert atmospheric nitrogen into a usable form for plants and other organisms. These bacteria can be free-living in the soil or form symbiotic relationships with certain plants, such as legumes (e.g., soybeans, clover) and some trees (e.g., acacia).

ii. **Ammonification:** Ammonification is the process by which decomposers, such as bacteria and fungi, break down organic nitrogen compounds found in dead plants, animals, and waste materials. During this process, organic nitrogen is converted into ammonium (NH_4^+), a form of inorganic nitrogen that can be readily used by plants.

iii. **Nitrification:** Nitrification is a two-step process that converts ammonium (NH_4^+) into nitrite (NO_2^-) and then into nitrate (NO_3^-). This conversion is carried out by nitrifying bacteria. Nitrate is the primary form of nitrogen taken up by plants and is crucial for their growth.

iv. **Assimilation:** Plants take up nitrate (NO_3^-) from the soil and assimilate it into their tissues. Nitrogen is then incorporated into plant proteins, DNA, and other organic molecules. Animals acquire nitrogen by consuming plants or other animals.

v. **Ammonium Release:** In some cases, instead of being assimilated by plants, ammonium (NH_4^+) can be released back into the soil by certain processes, such as root exudation or microbial activity. This ammonium can be reused by other plants or undergo nitrification again.

vi. **Denitrification:** Denitrification is the process in which specialized bacteria convert nitrate (NO_3^-) back into atmospheric nitrogen (N_2), completing the nitrogen cycle. This process occurs in oxygen-limited environments, such as waterlogged soils or sediments. Denitrification reduces the availability of nitrogen in the ecosystem.

Human activities, such as the use of nitrogen-based fertilizers in agriculture and industrial processes, can significantly impact the nitrogen cycle. Excessive use of fertilizers can lead to nitrogen runoff into water bodies, causing eutrophication and harmful algal blooms. Additionally, the combustion of fossil fuels releases nitrogen oxides into the atmosphere, contributing to air pollution and acid rain.

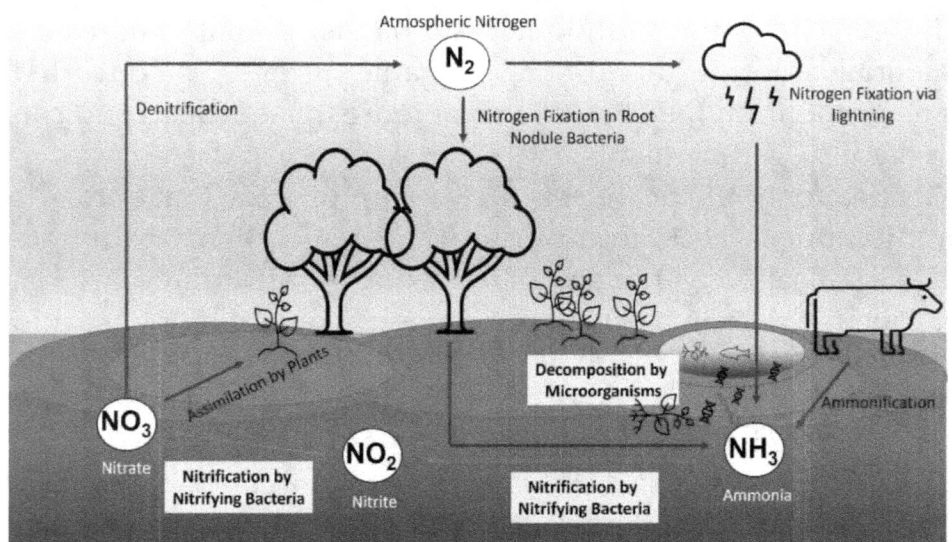

Oxygen Cycle

The oxygen cycle is the continuous movement and exchange of oxygen between the atmosphere, biosphere, and lithosphere. Oxygen is a vital element for the survival of most living organisms, as it plays a crucial role in cellular respiration and various metabolic processes. The oxygen cycle involves several interconnected processes that maintain the balance of oxygen in the environment.

i. **Photosynthesis:** Photosynthesis is the primary process that generates oxygen in the atmosphere. Green plants, algae, and some bacteria use sunlight, water, and carbon dioxide to produce glucose (energy-rich organic molecule) and release oxygen as a byproduct. This process takes place in chloroplasts, where the energy of sunlight is converted into chemical energy.

ii. **Respiration:** Respiration is the process by which organisms utilize oxygen to break down glucose and release energy for cellular activities. During respiration, oxygen is consumed, and carbon dioxide is produced as a waste product. Plants, animals, and microorganisms participate in respiration, contributing to the consumption and release of oxygen.

iii. **Decomposition:** Decomposition is the breakdown of organic matter by decomposers, such as bacteria and fungi. During decomposition, organic materials, including dead plants and animals, are broken down, and the organic compounds are metabolized. This process consumes oxygen and releases carbon dioxide, water, and other byproducts.

iv. **Combustion:** Combustion is the process of burning organic matter, such as fossil fuels (coal, oil, natural gas), wood, or biomass. During combustion, oxygen reacts with the organic compounds, releasing heat, carbon dioxide, and water vapor. Combustion is a significant source of carbon dioxide in the atmosphere and reduces the available oxygen.

v. **Atmospheric Exchange:** Oxygen and other gases undergo constant exchange between the atmosphere, oceans, and land. Diffusion and air movements, such as wind, facilitate the mixing of gases. The exchange occurs during gas exchange in plant leaves (where oxygen is released and carbon dioxide is absorbed) and through respiration by organisms.

Human activities can impact the oxygen cycle, particularly through deforestation and air pollution. Deforestation reduces the number of trees available for photosynthesis, leading to a decrease in oxygen production and an increase in carbon dioxide levels. Air pollution, caused by the burning of fossil fuels and industrial emissions, affects the balance of oxygen and other gases in the atmosphere.

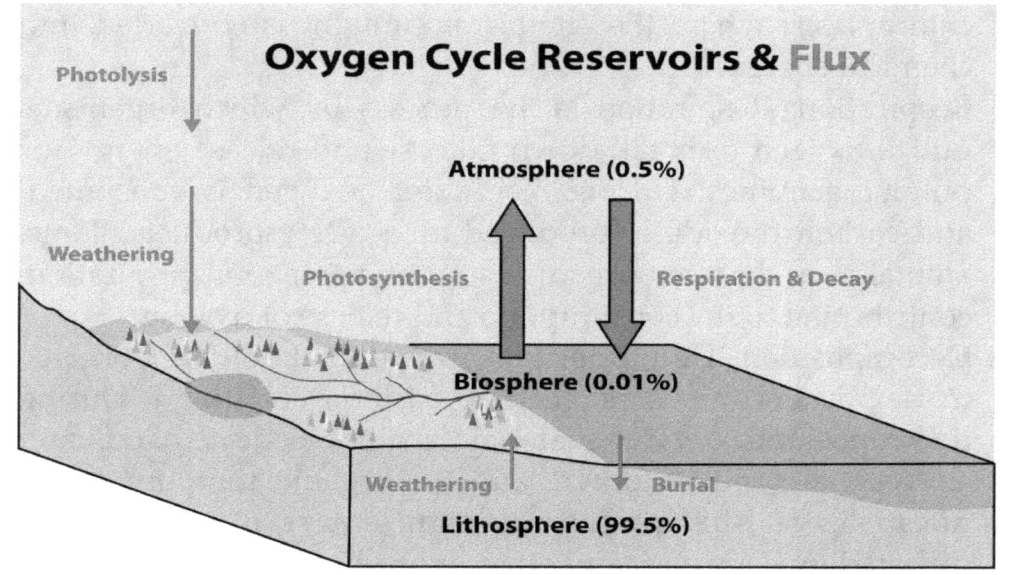

Water Cycle

The water cycle, also known as the hydrological cycle, is the continuous movement and exchange of water between the Earth's surface, the atmosphere, and back again. It involves various processes that contribute to the circulation and distribution of water in different forms. The water cycle is essential for maintaining the availability of freshwater for living organisms and supporting various ecological processes.

i. **Evaporation:** Evaporation is the process by which water changes from a liquid state to a gaseous state, primarily from the Earth's surface, including oceans, lakes, rivers, and even moist soil. The heat from the Sun provides the energy required to convert liquid water into water vapor. Evaporation is more rapid in areas with higher temperatures and strong winds.

ii. **Transpiration:** Transpiration is the release of water vapor from plants through their leaves and stems. During photosynthesis, plants absorb water from the soil through their roots. This water then travels up through the plant and is released into the atmosphere through tiny pores called stomata on the surface of leaves. Transpiration contributes significantly to the water vapor content in the atmosphere.

iii. **Condensation:** Condensation is the process by which water vapor in the atmosphere cools down and changes back into liquid water. As water vapor rises and encounters cooler air, it loses heat energy and forms tiny water droplets or ice crystals, which gather to form clouds. Condensation is facilitated by the presence of condensation nuclei, such as dust particles or aerosols, which provide surfaces for water vapor to condense onto.

iv. **Precipitation:** Precipitation occurs when condensed water droplets or ice crystals in clouds become large enough to fall from the atmosphere to the Earth's surface. Precipitation includes various forms, such as rain, snow, sleet, and hail. Precipitation can replenish water bodies, infiltrate into the soil, or flow over the land surface as runoff, eventually returning to the oceans or forming freshwater bodies.

v. **Runoff:** Runoff refers to the movement of water across the land surface, primarily as a result of gravity. When precipitation exceeds the capacity of the soil to absorb it, the excess water flows over the surface, forming streams, rivers, and eventually reaching lakes or oceans. Runoff plays a crucial role in shaping the landscape and transporting nutrients and sediments.

vi. **Infiltration:** Infiltration is the process by which water penetrates and moves into the soil from the land surface. Infiltrated water is stored in the soil as soil moisture, which can be taken up by plant roots, used by microorganisms, or percolate deeper into the ground, recharging groundwater reserves.

vii. **Groundwater:** Groundwater refers to the water that infiltrates through the soil and accumulates in porous rock layers, known as aquifers. Groundwater can be stored for long periods and serve as a source of water for wells, springs, and underground streams. It eventually resurfaces through springs or discharges into lakes, rivers, or the ocean.

viii. **Water Vapor Transport:** Water vapor in the atmosphere can be transported over long distances by winds. It moves in the form of invisible water vapor, condenses into clouds, and then releases precipitation in different regions. This transport helps distribute water resources across the Earth's surface.

The water cycle is a continuous and interconnected process that redistributes water across the planet, ensuring the availability of freshwater for various organisms and ecosystems. It plays a critical role in regulating global climate patterns, supporting plant growth, influencing weather systems, and shaping the Earth's surface.

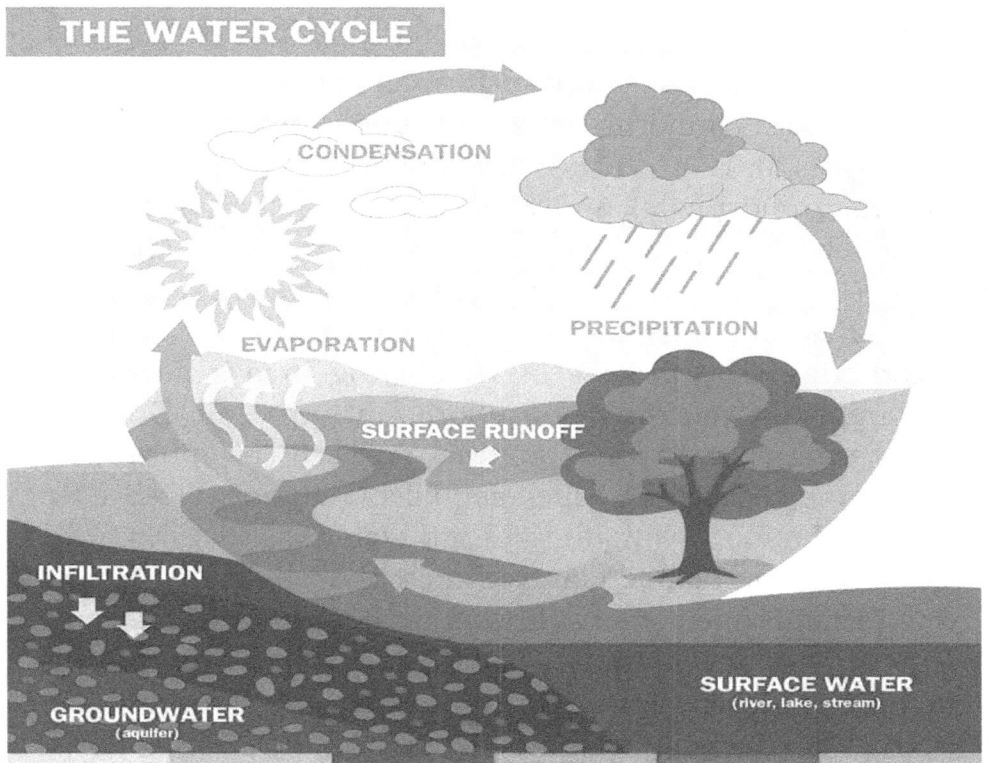

Phosphorus Cycle

The phosphorus cycle is a biogeochemical cycle that describes the movement and transformation of phosphorus in the environment. Phosphorus is an essential nutrient for living organisms, as it is a crucial component of DNA, RNA, ATP (adenosine triphosphate), and various other biological molecules. The phosphorus cycle involves several processes that cycle phosphorus between the lithosphere (rock and soil), hydrosphere (water bodies), and biosphere (living organisms).

i. **Weathering:** The process of weathering breaks down rocks and minerals, releasing phosphorus into the soil. Over time, weathering causes the gradual release of phosphorus-containing compounds, such as phosphate ions (PO_4^{3-}), from rocks and minerals into the soil and water.

ii. **Absorption by Plants:** Plants absorb phosphate ions from the soil through their roots. Phosphorus is essential for plant growth and development, playing a vital role in processes such as photosynthesis, energy transfer, and the synthesis of DNA and other cellular components.

iii. **Consumption by Animals:** When animals consume plants or other animals, they obtain phosphorus through their diet. Phosphorus is incorporated into the animal's tissues and used for various biological functions.

iv. **Decomposition:** When plants and animals die, their organic matter undergoes decomposition by bacteria and fungi. During decomposition, organic phosphorus compounds are broken down into inorganic phosphate ions, which can be released back into the soil.

v. **Sedimentation:** Phosphorus can enter aquatic ecosystems through runoff from the land. Phosphate ions in water can bind to sediment particles and settle at the bottom of lakes, rivers, and oceans over time. This process is called sedimentation and plays a significant role in the long-term storage of phosphorus.

vi. **Geological Uplift:** Through geological processes such as tectonic activity, the sedimentary rocks that contain phosphorus may be uplifted and exposed to weathering once again, restarting the cycle.

vii. **Leaching:** Phosphate ions in the soil can be subject to leaching, which is the movement of nutrients through water. Excessive rainfall or irrigation can cause phosphorus to be washed away from the soil, entering streams, rivers, and eventually reaching oceans.

viii. **Geological Timescale:** The process of geologic uplift and erosion occurs over long periods, and phosphorus can be locked in rock formations for millions of years before being released through weathering.

The phosphorus cycle differs from other biogeochemical cycles as it lacks a significant atmospheric component. Unlike carbon, nitrogen, and oxygen, phosphorus does not exist as a gas in the atmosphere. Instead, it primarily cycles through the soil, water, and organisms.

Introduction To Biology

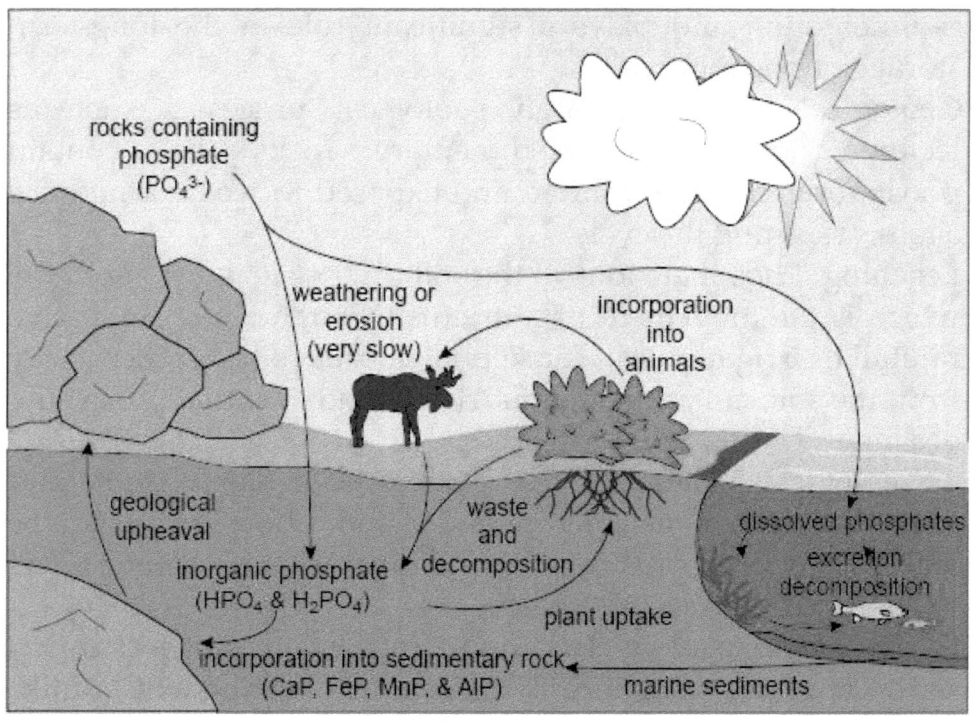

Other Nutrient Cycles: Besides carbon, nitrogen, and phosphorus, there are cycles for other essential nutrients like sulfur, potassium, calcium, and magnesium.

Nutrient cycling is vital for the functioning and productivity of ecosystems. It supports the growth and development of plants, which form the base of the food chain. Nutrient cycling helps maintain soil fertility, water quality, and the balance of essential elements in the environment. It also contributes to the regulation of climate by influencing carbon dioxide levels and the greenhouse effect.

Human activities can disrupt nutrient cycling in ecosystems. Practices such as deforestation, overuse of fertilizers, pollution, and improper waste management can alter nutrient levels, leading to nutrient imbalances, eutrophication, and degradation of ecosystem health. Sustainable land use practices and responsible nutrient management are essential to minimize these impacts and maintain the integrity of nutrient cycling processes.

■ Populations And Communities

In ecology, populations and communities are fundamental concepts that help us understand the dynamics and interactions of organisms within ecosystems. Let's explore these concepts in more detail:

1. **Populations:** A population refers to a group of individuals of the same species that live in the same geographic area and have the potential to interact and breed with one another. Populations are dynamic and can change in size, density, and composition over time.

a. **Population Size:** The total number of individuals in a population.
b. **Population Density:** The number of individuals per unit area or volume.
c. **Population Distribution:** The pattern of how individuals are spaced within a given area (e.g., clumped, uniform, random).
d. **Population Growth:** The change in population size over time, influenced by birth rates, death rates, immigration, and emigration.

Ecologists study populations to understand factors that affect their growth, such as availability of resources, predation, competition, disease, and environmental conditions. The study of populations provides insights into population dynamics, species interactions, and the processes that shape the structure of ecosystems.

2. **Communities:** A community refers to all the populations of different species living and interacting in a particular area or habitat. It represents the assemblage of various organisms that coexist and influence each other's distribution and abundance. Key characteristics of communities include:
   a. **Species Diversity:** The variety of different species present in a community.
   b. **Species Interactions:** The relationships between different species, such as competition, predation, mutualism, and symbiosis.
   c. **Trophic Structure:** The feeding relationships and energy flow within the community, including producers (plants), consumers (herbivores, carnivores, omnivores), and decomposers (fungi, bacteria).
   d. **Species Composition:** The specific species that make up the community and their relative abundance

Community ecology explores the interactions between species, the effects of species diversity on ecosystem functioning, and the processes that shape community structure. It examines topics such as competition for resources, predator-prey dynamics, mutualistic relationships, and the role of keystone species in maintaining community stability.

■ Conservation Biology

Conservation biology is a multidisciplinary field that focuses on the study and preservation of biodiversity, the sustainable management of natural resources, and the protection of ecosystems. It addresses the ongoing decline of species and habitats caused by human activities and seeks to develop strategies to mitigate these threats. Conservation biologists work towards understanding the value of biodiversity and implementing effective conservation measures to ensure the long-term survival of species and the functioning of ecosystems. Let's explore some key aspects of conservation biology:

1. **Biodiversity Conservation:** Conservation biology aims to protect the diversity of species and ecosystems on Earth. It recognizes the intrinsic value of biodiversity and the important ecological roles that species play. Conservation efforts involve identifying and prioritizing areas of high biodiversity, establishing protected areas, and implementing measures to prevent the loss of species and their habitats.

2. **Threatened Species and Endangered Species Recovery:** Conservation biologists study threatened and endangered species to understand their population dynamics, habitat requirements, and factors contributing to their decline. They work on developing recovery plans, implementing conservation actions, and monitoring the progress of species recovery programs. Techniques such as captive breeding, habitat restoration, and reintroduction programs are utilized to enhance the survival chances of endangered species.

3. **Habitat Conservation and Restoration:** Habitat loss and degradation are major threats to biodiversity. Conservation biologists work to identify and protect critical habitats, restore degraded ecosystems, and promote sustainable land and resource management practices. This involves studying the ecological requirements of species, assessing the impacts of habitat alteration, and developing conservation strategies that balance human needs with the preservation of natural habitats.

4. **Conservation Genetics:** Conservation biologists utilize genetic tools to assess the genetic diversity and population structure of endangered species. This information helps in identifying genetically unique populations, understanding the potential impacts of inbreeding and genetic drift, and developing strategies for maintaining genetic diversity within populations. Genetic techniques are also used in the identification of

illegally traded wildlife products and the enforcement of conservation laws.
5. **Ecosystem Services:** Conservation biology recognizes the critical role of ecosystems in providing essential services to human well-being. These services include the provision of clean water, air purification, climate regulation, soil fertility, and pollination. Conservation efforts aim to safeguard ecosystem services by protecting and restoring ecosystems, promoting sustainable resource use, and raising awareness about the value of nature.
6. **Conservation Policy and Management:** Conservation biology involves working with governments, organizations, and local communities to develop policies and management plans that promote conservation and sustainable development. This includes advocating for the implementation of protected areas, sustainable resource extraction practices, wildlife conservation laws, and international agreements to combat biodiversity loss.

VARIATION

■ Introduction To Variation

Variation refers to the differences or deviations that exist among individuals or groups within a population or species. It is a natural and inherent feature of living organisms and plays a significant role in shaping the diversity of life on Earth. Variation can occur at various levels, including genetic, phenotypic, and behavioral aspects, and is influenced by a combination of genetic and environmental factors. At the genetic level, variation refers to differences in the genetic material, specifically the DNA sequence, between individuals. This variation arises from genetic mutations, which are changes in the DNA sequence that can occur randomly or due to environmental factors. Mutations can introduce new genetic variants, leading to variations in traits and characteristics among individuals.

Phenotypic variation, on the other hand, refers to the observable differences in physical traits or characteristics among individuals. These traits can include morphological features (e.g., size, shape, color), physiological functions (e.g., metabolism, immune response), and behavioral patterns (e.g., feeding habits, mating rituals). Phenotypic variation is influenced by both genetic factors and environmental conditions. Genetic variation provides the foundation for phenotypic variation, but environmental factors such as nutrition, temperature, and exposure to stressors can also affect the expression of genes and contribute to phenotypic differences among individuals. Variation is essential for the process of evolution and adaptation. It provides the raw material for natural selection to act upon, as individuals with advantageous variations are more likely to survive, reproduce, and pass on their traits to future generations. Through the accumulation and inheritance of beneficial variations, populations can adapt to changing environmental conditions over time.

Variation also contributes to the diversity of species and ecosystems. It allows organisms to occupy different niches, adopt different strategies for survival and reproduction, and interact with their environment in various ways. Biodiversity, the variety of life on Earth, is a direct result of the immense variation that exists among organisms.

■ Types Of Variation

There are several types of variation that occur within populations or species. These types of variation can be classified based on their underlying causes and the aspects of an organism's characteristics that they affect. Here are some common types of variation:

1. **Genetic Variation:** Genetic variation refers to differences in the genetic makeup (DNA) between individuals. It is the result of genetic mutations, genetic recombination, and gene flow. Genetic variation is the raw material for evolution and provides the basis for natural selection to act upon. It leads to differences in traits and characteristics among individuals within a population.
2. **Phenotypic Variation:** Phenotypic variation refers to observable differences in physical traits or characteristics among individuals. These traits can include morphological features (e.g., height, eye color), physiological functions (e.g., metabolism, immune response), and behavioral patterns (e.g.,

foraging behavior, mating rituals). Phenotypic variation can be influenced by both genetic and environmental factors.

3. **Environmental Variation:** Environmental variation refers to differences in traits or characteristics among individuals that arise due to variations in environmental conditions. Environmental factors such as temperature, humidity, nutrient availability, and exposure to pollutants can influence an organism's phenotype. For example, plants growing in different light conditions may have variations in their leaf shape and size.

4. **Continuous Variation:** Continuous variation refers to a range of values or a spectrum of phenotypes that exist for a particular trait. In continuous variation, individuals show a gradual and smooth transition from one extreme of the trait to the other. Examples of continuous variation include height in humans or beak size in birds.

5. **Discontinuous Variation:** Discontinuous variation refers to distinct categories or discrete phenotypes that exist for a particular trait. In this type of variation, individuals fall into separate, non-overlapping groups or classes. Examples of discontinuous variation include blood types in humans or the presence or absence of wings in insects.

6. **Inherited Variation:** Inherited variation refers to the variation in traits that is passed down from parents to offspring through genetic inheritance. These traits are determined by the combination of genes inherited from both parents. Inherited variation is the basis for familial resemblances and the transmission of traits from one generation to the next.

7. **Environmental-Induced Variation:** Environmental-induced variation refers to changes in an organism's traits or characteristics that occur due to exposure to specific environmental factors during its lifetime. These changes are not inherited and do not affect the genetic makeup of the

organism. Examples of environmental-induced variation include changes in skin color due to sun exposure or changes in body weight due to diet and exercise.

■ Sources Of Variation

Genetic variation, the diversity in the genetic makeup of individuals within a population, arises from several sources. These sources contribute to the introduction of new genetic variants and the overall genetic diversity in a population. Here are the main sources of genetic variation:

1. **Genetic Mutations:** Genetic mutations are changes in the DNA sequence of an organism's genome. They can occur spontaneously during DNA replication or as a result of exposure to mutagens, which are agents that can increase the rate of mutations. Mutations can introduce new genetic variants into a population. They can be classified into different types, including point mutations (changes in a single nucleotide), insertions, deletions, and chromosomal rearrangements. Mutations can have various effects on an organism's phenotype, ranging from neutral to beneficial or detrimental.
2. **Genetic Recombination:** Genetic recombination occurs during sexual reproduction when genetic material from two parents is combined in offspring. It leads to the shuffling and recombination of genes, resulting in new combinations of alleles. Genetic recombination occurs through processes such as crossing over, where homologous chromosomes exchange genetic material, and independent assortment, where chromosomes align randomly during meiosis. Genetic recombination increases genetic diversity within a population and promotes the creation of new combinations of traits.

3. **Gene Flow:** Gene flow refers to the transfer of genetic material from one population to another through migration and interbreeding. When individuals from different populations mate, genetic information is exchanged, leading to the introduction of new alleles into the population. Gene flow can increase genetic diversity and homogenize the genetic composition of populations. It plays a crucial role in reducing genetic differentiation between populations and promoting the exchange of genetic variants.
4. **Genetic Drift:** Genetic drift refers to the random changes in allele frequencies that occur in small populations due to chance events. Genetic drift is more pronounced in small populations where random events, such as the death or reproduction of a few individuals, can have a significant impact on the genetic makeup of the population. Genetic drift can lead to the loss of genetic variants, a reduction in genetic diversity, and the fixation of alleles. It is more likely to occur in isolated populations or during population bottlenecks and founder events.
5. **Horizontal Gene Transfer:** Horizontal gene transfer (HGT) is the transfer of genetic material between organisms that are not parent and offspring. It is common in bacteria and other microorganisms but can also occur in higher organisms. HGT can transfer genes, including those involved in antibiotic resistance or other advantageous traits, between different species or even different kingdoms. HGT can lead to the rapid acquisition of new genetic variants and has played a significant role in the evolution of prokaryotes.

■ Factors Influencing Phenotypic Variation

Phenotypic variation, the observable differences in traits among individuals, is influenced by a combination of genetic and environmental factors. Here are some key factors that contribute to phenotypic variation:

1. **Genetic Factors:**
   i. **Genetic mutations:** Spontaneous changes in DNA sequence can lead to new genetic variants and phenotypic differences.
   ii. **Genetic recombination:** The shuffling and recombination of genes during sexual reproduction create new combinations of alleles, resulting in phenotypic variation.
   iii. **Gene expression:** Variation in gene expression levels and patterns can influence the production of proteins and other molecules, leading to phenotypic differences.
   iv. **Polygenic traits:** Many traits are influenced by multiple genes, with each gene contributing to a portion of the phenotype. Variations in these genes can result in phenotypic variation.

2. **Environmental Factors:**
   i. **Nutrient availability:** Differences in nutrient availability can affect growth, development, and overall phenotype of organisms.
   ii. **Temperature:** Organisms may exhibit phenotypic variations in response to different temperature regimes, such as changes in body size, metabolism, or coloration.
   iii. **Light exposure:** Light intensity and quality can influence traits such as plant growth and flowering time, as well as coloration in animals.
   iv. Climate and weather conditions: Environmental factors like humidity, precipitation, and seasonality can impact phenotypic traits and behaviors.
   v. **Resource availability and competition:** Variation in resource availability and competition for resources can drive differences in phenotypes among individuals.

vi. **Predation and predation pressure:** The presence or absence of predators can lead to phenotypic adaptations, such as camouflage or defensive traits.
3. **Developmental Factors:**
i. **Developmental plasticity:** Environmental cues during development can shape the phenotype of an organism, resulting in phenotypic variation.
ii. **Hormonal regulation:** Hormones play a crucial role in developmental processes and can influence phenotypic traits.
iii. **Epigenetic modifications:** Chemical modifications to DNA or associated proteins can influence gene expression patterns and result in phenotypic variation.

Interactions Between Genetic And Environmental Factors:

- **Gene-environment interactions:** Genetic variation can interact with environmental factors, leading to differential phenotypic responses.
- **Phenotypic plasticity:** Organisms may exhibit phenotypic plasticity, where they can adjust their phenotype in response to changing environmental conditions.

■ Significance Of Variation

Variation is of significant importance in biological systems and plays a crucial role in various aspects of life. Here are some key significances of variation:
1. **Evolutionary Adaptation:** Variation provides the raw material for natural selection, the driving force behind evolutionary change. Individuals with advantageous variations are more likely to survive and reproduce, passing on their beneficial traits to future generations. Genetic and phenotypic variations allow populations to adapt to changing environmental

conditions, increasing their chances of survival and successful reproduction. Variation contributes to the diversification of species over time, leading to the evolution of new forms, functions, and ecological niches.
2. **Biodiversity and Ecological Stability:** Variation within and between species is essential for maintaining biodiversity, which refers to the variety of living organisms and ecosystems on Earth. Biodiversity ensures ecological stability by promoting resilience and resistance to environmental changes, such as climate fluctuations or the introduction of new predators or diseases. Species with high genetic and phenotypic variation have a greater capacity to respond to environmental challenges, enhancing the overall health and stability of ecosystems.
3. **Disease Resistance and Health:** Genetic variation plays a crucial role in determining an individual's susceptibility to diseases. Variations in immune system genes, for example, can impact an individual's ability to combat pathogens. Genetic diversity within populations reduces the risk of widespread disease outbreaks by increasing the likelihood that some individuals will possess resistance to specific pathogens. Understanding genetic variation can aid in the development of personalized medicine, as different individuals may respond differently to treatments based on their genetic makeup.
4. **Agricultural and Food Security:** Variation in crop plants and livestock is essential for breeding programs aimed at improving agricultural productivity, disease resistance, and nutritional quality. Genetic diversity in agricultural systems ensures that there are diverse genetic resources available to cope with changing environmental conditions, pests, and diseases. Preservation of wild relatives of crop plants and

genetic diversity in seed banks is crucial for maintaining long-term food security.
5. **Conservation and Ecosystem Management:** Conservation efforts aim to preserve genetic and phenotypic variation within species, as it represents the basis for adaptation and evolutionary potential. Understanding variation is essential for effective conservation strategies, such as captive breeding programs, habitat restoration, and the identification of genetically unique or vulnerable populations. Variation within ecosystems contributes to ecological resilience, allowing ecosystems to better withstand disturbances and maintain their essential functions and services.

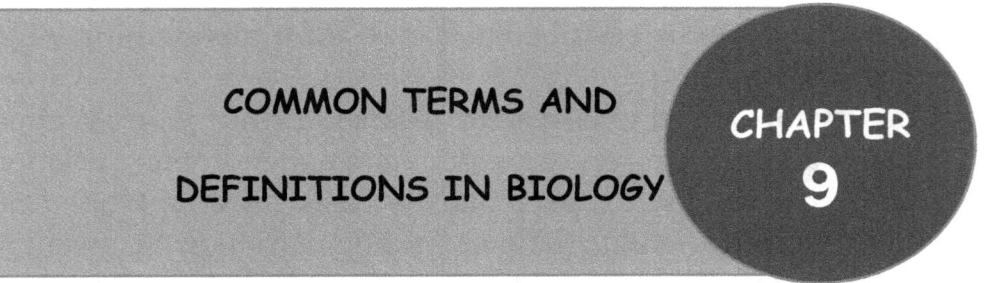

COMMON TERMS AND DEFINITIONS IN BIOLOGY

CHAPTER 9

1. **Abiotic:** Refers to non-living factors or components in an ecosystem, such as temperature, sunlight, soil, and water.
2. **Allele:** Variant forms of a gene that arise through mutation and are located at the same position on a specific chromosome.
3. **Antibody**: A Y-shaped protein produced by the immune system in response to the presence of antigens, which helps neutralize or eliminate pathogens.
4. **Asexual reproduction:** Reproduction that involves only one parent and produces offspring that are genetically identical to the parent.
5. **ATP (Adenosine triphosphate):** A molecule that stores and supplies energy for cellular processes in organisms.
6. **Biodiversity:** The variety of life forms, including species diversity, genetic diversity, and ecosystem diversity, in a particular habitat or on Earth as a whole.
7. **Biome:** A large-scale community of plants and animals that occupies a specific type of habitat characterized by its climate and vegetation.
8. **Biotechnology:** The use of living organisms or their components to produce useful products or perform specific tasks for human benefit.
9. **Cell:** The basic structural and functional unit of all living organisms, capable of carrying out essential processes of life.
10. **Chromosome:** A thread-like structure composed of DNA and proteins, carrying genetic information in the form of genes.

11. **Carbohydrate:** A biomolecule composed of carbon, hydrogen, and oxygen, serving as a primary source of energy and providing structural support in cells.
12. **Cell membrane:** A thin, flexible barrier that surrounds the cell, regulating the movement of substances in and out of the cell.
13. **Chloroplast:** An organelle found in plant cells and some protists, responsible for photosynthesis and containing chlorophyll.
14. **Community:** A group of different populations of organisms living and interacting together in the same habitat.
15. **Cytoplasm:** The jelly-like fluid inside the cell that surrounds the organelles and is involved in various cellular processes.
16. **DNA (Deoxyribonucleic acid):** A molecule that carries the genetic instructions used in the development and functioning of all known living organisms.
17. **Dominant:** A genetic trait or allele that is expressed or observable in an individual even if it is present in only one copy.
18. **Ecosystem:** A biological community of interacting organisms and their physical environment, including both living and non-living components.
19. **Endoplasmic reticulum:** An organelle involved in the synthesis, folding, modification, and transport of proteins and lipids within the cell.
20. **Enzyme:** A protein that acts as a biological catalyst, speeding up chemical reactions in living organisms.
21. **Evolution:** The process of change in inherited characteristics of species over successive generations, driven by mechanisms such as natural selection, mutation, and genetic drift.
22. **Excretion:** The process by which metabolic waste products and toxins are eliminated from an organism, typically through organs like kidneys, lungs, or skin.

23. **Extinction:** The complete disappearance of a species from Earth, either due to natural processes or human activities.
24. **Exponential growth:** A rapid and continuous increase in the population size of organisms, where the growth rate is proportional to the current population size.
25. **Eyepiece:** The lens or group of lenses at the top of a microscope through which the observer looks to view the magnified image.
26. **Fermentation:** The metabolic process that converts sugar into acids, gases, or alcohol in the absence of oxygen, often carried out by microorganisms.
27. **Fertilization:** The process by which a sperm cell fuses with an egg cell to form a zygote, initiating the development of a new organism.
28. **Food chain:** A sequence of organisms in which each organism serves as a source of energy for the next, depicting the transfer of energy and nutrients in an ecosystem.
29. **Fungi:** A kingdom of eukaryotic organisms that includes mushrooms, yeasts, and molds, known for their ability to decompose organic matter and form symbiotic associations.
30. **Gene:** A hereditary unit of DNA that contains the instructions for building and maintaining an organism, transmitting traits from parents to offspring.
31. **Gene expression:** The process by which the information in a gene is used to synthesize a functional gene product, such as a protein or RNA molecule.
32. **Genotype:** The genetic makeup of an organism, represented by the specific combination of alleles present in its DNA.
33. **Germ cell:** The specialized cell that gives rise to gametes (sperm or eggs) during the process of sexual reproduction.
34. **Golgi apparatus:** An organelle involved in the modification, sorting, and packaging of proteins for transport within or outside the cell.

35. **Growth hormone:** A hormone produced by the pituitary gland that stimulates growth and development in organisms.
36. **Habitat:** The specific environment or place where an organism or a population of organisms naturally lives and obtains the resources it needs to survive.
37. **Heterozygous:** Having two different alleles for a particular gene, resulting in a hybrid genotype.
38. **Homeostasis:** The ability of an organism to maintain stable internal conditions, such as body temperature, pH levels, and nutrient balance, despite external fluctuations.
39. **Hormone:** A chemical substance produced by endocrine glands that acts as a signaling molecule, regulating various physiological processes in the body.
40. **Hybridization:** The breeding or crossing of individuals from two different species, varieties, or populations to produce offspring with mixed traits.
41. **Immunity:** The ability of an organism to resist and defend against pathogens or foreign substances through the actions of the immune system.
42. **Inheritance:** The passing of traits or characteristics from parents to offspring through genetic information carried by genes.
43. **Inorganic:** Referring to substances or compounds that do not contain carbon and are not derived from living organisms.
44. **Insulin:** A hormone produced by the pancreas that regulates blood glucose levels by facilitating the uptake and storage of glucose by cells.
45. **Interphase:** The longest phase of the cell cycle, during which a cell grows, carries out normal functions, and prepares for division.
46. **Joint:** A structure where two or more bones meet, allowing movement and providing support to the skeletal system.

Introduction To Biology

47. **Jumping genes:** Also known as transposable elements or transposons, these are genetic elements that can move or "jump" within the genome, influencing genetic variation and evolution.
48. **Jurkat cells:** A line of immortalized human T lymphocyte cells that is commonly used in scientific research and studies related to immunology and cancer.
49. **Juvenile:** Referring to the early stage of development or growth in an organism, typically after the embryonic stage and before reaching adulthood.
50. **Juxtaglomerular cells:** Specialized cells in the kidney that are involved in regulating blood pressure and fluid balance by producing and secreting renin.
51. **Karyotype:** The number, size, and shape of chromosomes in an individual or species, typically displayed in a standardized arrangement.
52. **Kingdom:** The highest taxonomic rank in the classification of organisms, grouping organisms into five major categories: Animalia, Plantae, Fungi, Protista, and Monera (now divided into Bacteria and Archaea).
53. **Krebs cycle:** Also known as the citric acid cycle or tricarboxylic acid cycle, it is a series of chemical reactions that occur in the mitochondria and are involved in the production of energy-rich molecules during cellular respiration.
54. **Kinesis:** A type of non-directional movement or response exhibited by organisms in response to a stimulus, characterized by changes in activity or speed of movement.
55. **Kinetochore:** A protein structure located on the centromere of a chromosome that plays a crucial role in the attachment of the chromosome to the spindle fibers during cell division.
56. **Lysosome:** An organelle within eukaryotic cells that contains digestive enzymes responsible for breaking down waste materials, cellular debris, and macromolecules.

57. **Ligament:** A tough band of fibrous connective tissue that connects bones to other bones, providing stability and facilitating movement in joints.
58. **Lipid:** A class of biomolecules that include fats, oils, phospholipids, and steroids, which serve various functions such as energy storage, insulation, and component of cell membranes.
59. **Locus:** The specific position or location of a gene on a chromosome.
60. **Lytic cycle:** A viral replication cycle in which the virus infects a host cell, replicates within it, and ultimately causes the host cell to burst (lyse), releasing new virus particles.
61. **Meiosis:** A specialized cell division process that occurs in sexually reproducing organisms, resulting in the formation of haploid gametes (sperm and eggs) with half the number of chromosomes.
62. **Mitochondria:** Membrane-bound organelles found in eukaryotic cells that are often referred to as the "powerhouses" of the cell because they generate energy in the form of ATP through cellular respiration.
63. **Mutation:** A change or alteration in the DNA sequence of a gene, which can result in genetic variation and contribute to evolution.
64. **Mycelium:** The vegetative part of a fungus, consisting of a network of branching hyphae that infiltrate the substrate and absorb nutrients.
65. **Natural selection:** The process by which certain heritable traits become more or less common in a population over successive generations due to differential reproductive success based on fitness.
66. **Nucleotide:** The building blocks of nucleic acids (DNA and RNA), composed of a sugar molecule, a phosphate group, and a nitrogenous base.

Introduction To Biology

67. **Nucleus:** The membrane-bound organelle found in eukaryotic cells that contains the genetic material (DNA) and controls cellular activities.
68. **Nutrient:** A substance or compound required by organisms for growth, development, and maintenance of vital functions.
69. **Niche:** The specific role or position occupied by an organism within an ecosystem, including its interactions with other organisms and its utilization of resources.
70. **Nervous system:** The complex network of specialized cells (neurons) and tissues that transmit signals between different parts of an organism, coordinating its responses to the environment.
71. **Organ:** A structure composed of different tissues working together to perform specific functions in an organism, such as the heart, lungs, or liver.
72. **Osmosis:** The movement of solvent molecules (usually water) across a selectively permeable membrane from an area of lower solute concentration to an area of higher solute concentration.
73. **Organism:** A living individual, typically composed of one or more cells, capable of carrying out essential life processes such as growth, reproduction, and response to stimuli.
74. **Ovary:** In female organisms, the reproductive organ that produces eggs (ova) and secretes hormones.
75. **Ozone layer:** A region of the Earth's stratosphere that contains a high concentration of ozone (O_3) molecules, protecting the planet by absorbing most of the Sun's harmful ultraviolet (UV) radiation.
76. **Photosynthesis:** The process by which green plants, algae, and some bacteria convert sunlight, carbon dioxide, and water into glucose (a form of chemical energy) and release oxygen as a byproduct.

Introduction To Biology

77. **Population**: A group of individuals of the same species that inhabit a specific geographic area and can interbreed with one another.
78. **Prokaryote**: A type of cell that lacks a nucleus and membrane-bound organelles. Prokaryotes include bacteria and archaea.
79. **Protein**: A large biomolecule composed of amino acids linked together by peptide bonds. Proteins perform a wide range of functions in cells, including structural support, enzymatic activity, and cellular signaling.
80. **Punnett square**: A grid-based diagram used to predict the possible genotypes and phenotypes of offspring in a genetic cross, based on the known genotypes of the parents.
81. **Quadrat**: A square or rectangular frame used in ecological studies to estimate population densities or assess the distribution of organisms within a defined area.
82. **Quorum sensing**: A process by which bacteria and some other microorganisms communicate and coordinate their behavior based on the local population density through the release and detection of signaling molecules.
83. **Quaternary structure**: The specific arrangement and association of multiple protein subunits to form a functional protein complex.
84. **Quantitative trait**: A heritable characteristic or trait that can be measured and quantified, such as height, weight, or blood pressure.
85. **Quiescent**: In a state of dormancy or temporary inactivity, often used to describe cells that are not actively dividing.
86. **Respiration**: The biochemical process by which cells obtain energy from organic molecules, typically using oxygen and producing carbon dioxide as a byproduct.
87. **Ribosome**: A cellular structure composed of RNA and proteins, responsible for protein synthesis by translating messenger RNA (mRNA) into specific amino acid sequences.

88. **RNA (Ribonucleic acid):** A nucleic acid molecule involved in various cellular processes, including gene expression and protein synthesis.
89. **Replication:** The process by which DNA is copied or duplicated, resulting in the formation of two identical DNA molecules.
90. **Recessive:** A genetic trait or allele that is expressed or observable only when present in two copies (homozygous), as it is masked by a dominant allele in heterozygotes.
91. **Selective breeding:** The intentional breeding of organisms with desirable traits to produce offspring with those traits, leading to the genetic improvement of populations over generations.
92. **Symbiosis:** A close and long-term interaction between two or more different species, often benefiting one or both of the participating organisms.
93. **Spore:** A reproductive structure produced by certain organisms, such as fungi, plants, and some bacteria, that is capable of giving rise to a new individual.
94. **Stimulus:** A detectable change in the internal or external environment that evokes a response or reaction in an organism.
95. **Synapse:** A specialized junction between two neurons or between a neuron and a target cell, where signals are transmitted through chemical or electrical signaling.
96. **Taxonomy:** The science of classifying and categorizing organisms into hierarchical groups based on their evolutionary relationships and shared characteristics.
97. **Tissue:** A group or collection of similar cells that work together to perform a specific function within an organism.
98. **Transcription:** The process of synthesizing RNA from a DNA template, resulting in the production of messenger RNA (mRNA) molecules.

Introduction To Biology

99. **Trophic level:** The position of an organism in a food chain or food web, representing its feeding relationship and energy transfer within an ecosystem.
100. **Transgenic:** Referring to an organism that has been genetically modified by introducing foreign genes or DNA sequences from another species.
101. **Ubiquitin:** A small protein that plays a crucial role in marking other proteins for degradation by the proteasome, a cellular protein disposal system.
102. **Urea:** A nitrogenous waste product formed in the liver from the breakdown of proteins and amino acids. It is excreted by the kidneys in mammals and some other organisms.
103. **Ultrastructure:** The detailed structure of cells and tissues as revealed by electron microscopy, providing high-resolution images of cellular components and organelles.
104. **Uracil:** One of the four nitrogenous bases found in RNA, it pairs with adenine during RNA synthesis and does not occur in DNA.
105. **Unicellular:** Referring to organisms that consist of a single cell, as opposed to multicellular organisms.
106. **Vaccine:** A biological preparation containing weakened or inactivated pathogens (or parts of them) that stimulates the immune system to produce a protective response against specific diseases.
107. **Vector:** In biology, a vector can refer to an organism (such as an insect) that transmits a disease-causing agent from one host to another or a DNA molecule used to transfer genetic material into a host cell.
108. **Virus:** A microscopic infectious agent that consists of genetic material (DNA or RNA) enclosed in a protein coat. Viruses can infect a wide range of organisms and cause various diseases.

109. **Vesicle:** A small, membrane-bound sac that transports or stores substances within cells, facilitating intracellular transport and communication.
110. **Variation:** The differences or range of variation observed within a population or among different populations of a species, often resulting from genetic, environmental, or developmental factors.
111. **Watson-Crick model:** The double helix structure of DNA proposed by James Watson and Francis Crick in 1953, which elucidates the complementary base pairing and overall molecular organization of DNA.
112. **Xerophyte:** A type of plant that is adapted to arid or dry environments, characterized by specialized adaptations to conserve water, such as reduced leaves, thick cuticles, or succulent stems.
113. **Y-linked inheritance:** The pattern of inheritance observed for genes located on the Y chromosome, which is passed down exclusively from father to son.
114. **Zygote:** The fertilized egg formed by the fusion of a sperm cell and an egg cell during sexual reproduction, which develops into an embryo.
115. **Zoonosis:** A disease or infection that can be transmitted between animals and humans, either directly or through intermediate hosts.
116. **Zone of Inhibition:** The area surrounding an antimicrobial agent (such as an antibiotic) on a culture plate where bacterial growth is inhibited, indicating the effectiveness of the agent against the specific bacteria.
117. **Zoology:** The branch of biology that focuses on the study of animals, including their classification, anatomy, behavior, and distribution.

Introduction To Biology

118. **Zygomorphic:** Referring to a flower or organism that exhibits bilateral symmetry, meaning it can be divided into two equal halves in only one plane.

CONGRATULATIONS UPON COMPLETION OF THIS TEXTBOOK!

www.ingramcontent.com/pod-product-compliance
Lightning Source LLC
Chambersburg PA
CBHW071447220526
45472CB00003B/709